T0164901

HEXAFLEXAGONS, PROBABILITY PARADOXES, AND THE TOWER OF HANOI

For 25 of his 90 years, Martin Gardner wrote "Mathematical Games and Recreations," a monthly column for *Scientific American* magazine. These columns have inspired hundreds of thousands of readers to delve more deeply into the large world of mathematics. He has also made significant contributions to magic, philosophy, debunking pseudoscience, and children's literature. He has produced more than 60 books, including many best sellers, most of which are still in print. His *Annotated Alice* has sold more than a million copies. He continues to write a regular column for the *Skeptical Inquirer* magazine. (The photograph is of the author at the time of the first edition.)

THE NEW MARTIN GARDNER MATHEMATICAL LIBRARY

Editorial Board

Donald J. Albers,   *Menlo College*
Gerald L. Alexanderson,   *Santa Clara University*
John H. Conway, F. R. S.,   *Princeton University*
Richard K. Guy,   *University of Calgary*
Harold R. Jacobs
Donald E. Knuth,   *Stanford University*
Peter L. Renz

From 1957 through 1986 Martin Gardner wrote the "Mathematical Games" columns for *Scientific American* that are the basis for these books. *Scientific American* editor Dennis Flanagan noted that this column contributed substantially to the success of the magazine. The exchanges between Martin Gardner and his readers gave life to these columns and books. These exchanges have continued and the impact of the columns and books has grown. These new editions give Martin Gardner the chance to bring readers up to date on newer twists on old puzzles and games, on new explanations and proofs, and on links to recent developments and discoveries. Illustrations have been added and existing ones improved, and the bibliographies have been greatly expanded throughout.

1. Hexaflexagons, Probability Paradoxes, and the Tower of Hanoi: Martin Gardner's First Book of Mathematical Puzzles and Games
2. Origami, Eleusis, and the Soma Cube: Martin Gardner's Mathematical Diversions
3. Sphere Packing, Lewis Carroll, and Reversi: Martin Gardner's New Mathematical Diversions
4. Knots and Borromean Rings, Rep-Tiles, and Eight Queens: Martin Gardner's Unexpected Hanging
5. Klein Bottles, Op-Art, and Sliding-Block Puzzles: More of Martin Gardner's Mathematical Games
6. Sprouts, Hypercubes, and Superellipses: Martin Gardner's Mathematical Carnival

Hexaflexagons, Probability Paradoxes, and the Tower of Hanoi

MARTIN GARDNER'S FIRST BOOK OF MATHEMATICAL PUZZLES AND GAMES

Martin Gardner

The Mathematical Association of America

CAMBRIDGE
UNIVERSITY PRESS

CAMBRIDGE
UNIVERSITY PRESS

32 Avenue of the Americas, New York NY 10013-2473, USA

Cambridge University Press is part of the University of Cambridge.

It furthers the University's mission by disseminating knowledge in the pursuit of education, learning, and research at the highest international levels of excellence.

www.cambridge.org
Information on this title: www.cambridge.org/9780521735254

© Mathematical Association of America 2008

This publication is in copyright. Subject to statutory exception and to the provisions of relevant collective licensing agreements, no reproduction of any part may take place without the written permission of Cambridge University Press.

First edition published as The SCIENTIFIC AMERICAN Book of Mathematical Puzzles & Diversions, Simon and Schuster, 1959
First published 2008
7th printing 2013

A catalog record for this publication is available from the British Library.

Library of Congress Cataloging in Publication data
Gardner, Martin, 1914–
Hexaflexagons, probability paradoxes, and the Tower of Hanoi : Martin Gardner's first book of mathematical puzzles and games / Martin Gardner.
  p. cm. – (The new Martin Gardner mathematical library)
Includes bibliographical references and index.
ISBN 978-0-521-75615-0 (hardback)
1. Mathematical recreations. I. Title. II. Series.
QA95.G247 2008
793.74 – dc22   2008012533

ISBN 978-0-521-75615-0 Hardback
ISBN 978-0-521-73525-4 Paperback

Cambridge University Press has no responsibility for the persistence or accuracy of URLs for external or third-party Internet Web sites referred to in this publication, and does not guarantee that any content on such Web sites is, or will remain, accurate or appropriate.

Contents

Acknowledgments

MARTIN GARDNER thanks Scientific American for allowing reuse of material from his columns in that magazine, material Copyright © 1956 (Chapter 1), and 1957 (Chapters 2–13), and 1958 (Chapters 14–16) by Scientific American, Inc. He also thanks the artists who contributed to the success of these columns and books for allowing reuse of their work: James D. Egelson (via heirs Jan and Nicholas Egleson), Irving Geis (via heir Sandy Geis), Harold Jacobs, Amy Kasai, and Bunji Tagawa (via Donald Garber for the Tagawa Estate). Artists names are cited where these were known. All rights other than use in connection with these materials lie with the original artists.

Photograph in Figure 48 is courtesy of the Museum of Fine Arts, Boston, 2008. Used by permission.

Introduction to the First Edition

THE ELEMENT OF PLAY, which makes recreational mathematics recreational, may take many forms: a puzzle to be solved, a competitive game, a magic trick, paradox, fallacy, or simply mathematics with any sort of curious or amusing fillip. Are these examples of pure or applied mathematics? It is hard to say. In one sense recreational mathematics is pure mathematics, uncontaminated by utility. In another sense it is applied mathematics, for it meets the universal human need for play.

Perhaps this need for play is behind even pure mathematics. There is not much difference between the delight a novice experiences in cracking a clever brain teaser and the delight a mathematician experiences in mastering a more advanced problem. Both look on beauty bare – that clean, sharply defined, mysterious, entrancing order that underlies all structure. It is not surprising, therefore, that it is often difficult to distinguish pure from recreational mathematics. The four-color map theorem, for example, is an important theorem in topology, yet discussions of the theorem will be found in many recreational volumes. No one can deny that paper flexagons, the subject of this book's opening chapter, are enormously entertaining toys; yet an analysis of their structure takes one quickly into advanced group theory, and articles on flexagons have appeared in technical mathematical journals.

Creative mathematicians are seldom ashamed of their interest in recreational mathematics. Topology had its origin in Euler's analysis of a puzzle about crossing bridges. Leibniz devoted considerable time to the study of a peg-jumping puzzle that recently enjoyed its latest revival under the trade name of Test Your High-Q. David

Hilbert, the great German mathematician, proved one of the basic theorems in the field of dissection puzzles. Alan Turing, a pioneer in modern computer theory, discussed Sam Loyd's 15-puzzle (here described in Chapter 9) in an article on solvable and unsolvable problems. I have been told by Piet Hein (whose game of Hex is the subject of Chapter 8) that when he visited Albert Einstein he found a section of Einstein's bookshelf devoted to books on recreational mathematics. The interest of those great minds in mathematical play is not hard to understand, for the creative thought bestowed on such trivial topics is of a piece with the type of thinking that leads to mathematical and scientific discovery. What is mathematics, after all, except the solving of puzzles? And what is science if it is not a systematic effort to get better and better answers to puzzles posed by nature?

The pedagogic value of recreational mathematics is now widely recognized. One finds an increasing emphasis on it in magazines published for mathematics teachers, and in the newer textbooks, especially those written from the "modern" point of view. *Introduction to Finite Mathematics*, for example, by J. G. Kemeny, J. Laurie Snell, and Gerald L. Thompson, is livened by much recreational material. These items hook a student's interest as little else can. The high school mathematics teacher who reprimands two students for playing a surreptitious game of ticktacktoe instead of listening to the lecture might well pause and ask: "Is this game more interesting mathematically to these students than what I am telling them?" In fact, a classroom discussion of ticktacktoe is not a bad introduction to several branches of modern mathematics.

In an article on "The Psychology of Puzzle Crazes" (*Nineteenth Century Magazine*, December 1926) the great English puzzlist Henry Ernest Dudeney made two complaints. The literature of recreational mathematics, he said, is enormously repetitive, and the lack of an adequate bibliography forces enthusiasts to waste time in devising problems that have been devised long before. I am happy to report that the need for such a bibliography has at last been met. Professor William L. Schaaf, of Brooklyn College, compiled an excellent four-volume bibliography, titled *Recreational Mathematics*, which can be obtained from the National Council of Teachers of Mathematics. As to Dudeney's other complaint, I fear that it still applies to current books in the field, including this one, but I think readers will discover here more than the usual portion of fresh material that has not previously found its way between book covers.

I would like to thank Gerard Piel, publisher of *Scientific American*, and Dennis Flanagan, editor, for the privilege of appearing regularly in the distinguished company of their contributors, and for permission to reprint my efforts in the present volume. And I am grateful also to the thousands of readers, from all parts of the world, who have taken the trouble to call my attention to mistakes (alas too frequent) and to make valuable suggestions. In some cases this welcome feedback has been incorporated into the articles themselves, but in most cases it is pulled together in an addendum at the end of each chapter. The answers to problems (where necessary) also appear at the end of the chapter. A bibliography of selected references for further reading will be found at the close of the book.

And I must not fail to thank my wife, not only for competent and fairly cheerful proofreading, but also for her patience during those trying moments of mathematical meditation when I do not hear what she is saying.

Martin Gardner
Dobbs Ferry, New York, 1959

Preface to the Second Edition

FOR MORE THAN TWENTY-FIVE YEARS I wrote a monthly column on recreational mathematics for *Scientific American*. Those columns have been reprinted in fifteen books. In 2005, when the Mathematical Association of America (MAA) put all fifteen on a CD, type was not reset. This severely limited what I could add to update the columns and expand bibliographies. Because Cambridge University Press is resetting type, I am now happily free to add as much fresh material as I please.

I am indebted to Don Albers and to Peter Renz for initiating the MAA's joint venture with Cambridge to produce a uniform set of all the *Scientific American* books and to Elwyn Berlekamp for support of preparation of the manuscript. I am equally indebted to my many readers, both professional and amateur mathematicians, for supplying so much new material for my columns.

I'm not a creative mathematician. I am a journalist who loves math and who enjoys writing about what the real mathematicians discover. Note that I say "discover." I'm an unabashed Platonist who believes, with all the great mathematicians past and present, that mathematical truth is independent of human cultures. It is as firmly "out there," in its own strange and mysterious abstract realm, as the stars are out there as material structures not made by us.

Martin Gardner
Norman, Oklahoma, 2008

CHAPTER ONE

Hexaflexagons

FLEXAGONS ARE PAPER POLYGONS, folded from straight or crooked strips of paper, which have the fascinating property of changing their faces when they are "flexed." Had it not been for the trivial circumstance that British and American notebook paper are not the same size, flexagons might still be undiscovered, and a number of top-flight mathematicians would have been denied the pleasure of analyzing their curious structures.

It all began in the fall of 1939. Arthur H. Stone, a 23-year-old graduate student from England, in residence at Princeton University on a mathematics fellowship, had just trimmed an inch from his American notebook sheets to make them fit his English binder. For amusement he began to fold the trimmed-off strips of paper in various ways, and one of the figures he made turned out to be particularly intriguing. He had folded the strip diagonally at three places and joined the ends so that it made a hexagon (see Figure 1). When he pinched two adjacent triangles together and pushed the opposite corner of the hexagon toward the center, the hexagon would open out again, like a budding flower, and show a completely new face. If, for instance, the top and bottom faces of the original hexagon were painted different colors, the new faces would come up blank and one of the colored faces would disappear!

This structure, the first flexagon to be discovered, has three faces. Stone did some thinking about it overnight and, on the following day, confirmed his belief (arrived at by pure cerebration) that a more complicated hexagonal model could be folded with six faces instead of only three. At this point, Stone found the structure so interesting that he showed his paper models to friends in the graduate school.

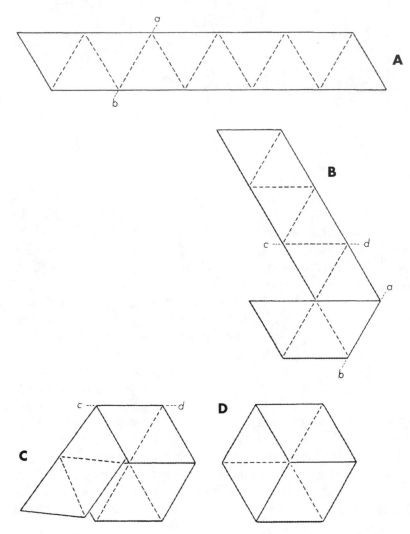

Figure 1. Trihexaflexagon is constructed by cutting a strip of paper so that it may be marked off in 10 equilateral triangles (A). The strip is folded backward along the line *ab* and turned over (B). It is then folded backward again along the line *cd* and the next-to-the-last-triangle placed on top of the first (C). The last triangle is now folded backward and glued to the other side of the first (D). Fairly stiff paper at least an inch and a half wide is recommended. (Artist: Bunji Tagawa)

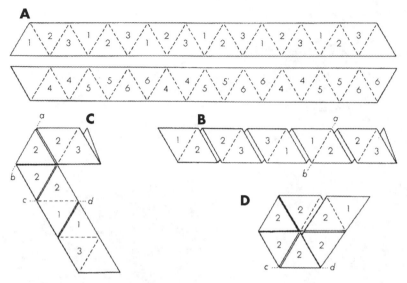

Figure 2. Hexahexaflexagon is constructed by cutting a strip of paper so that it may be marked off in 19 triangles (A). The triangles on one side are numbered 1, 2, and 3; the triangles on the other, 4, 5, and 6. A similar pattern of colors or geometrical figures may also be used. The strip of paper is then folded as shown (B–D). It can be flexed to show six different faces. (Artist: Bunji Tagawa)

Soon "flexagons" were appearing in profusion at the lunch and dinner tables. The Flexagon Committee was organized to probe further into the mysteries of flexigation. The other committee members besides Stone were Bryant Tuckerman, a graduate student of mathematics; Richard P. Feynman, a graduate student in physics; and John W. Tukey, a young mathematics instructor.

The models were named hexaflexagons – *hexa* for their hexagonal form and *flexagon* for their ability to flex. Stone's first model is a trihexaflexagon (*tri* for the three different faces that can be brought into view); his elegant second structure is a hexahexaflexagon (for its six faces).

To make a hexahexaflexagon, you start with a strip of paper (the tape used in adding machines serves admirably), which is divided into 19 equilateral triangles (see Figure 2A). You number the triangles on one side of the strip 1, 2, and 3, leaving the nineteenth triangle blank, as shown in the drawing. On the opposite side the triangles are numbered 4, 5, and 6, according to the scheme shown.

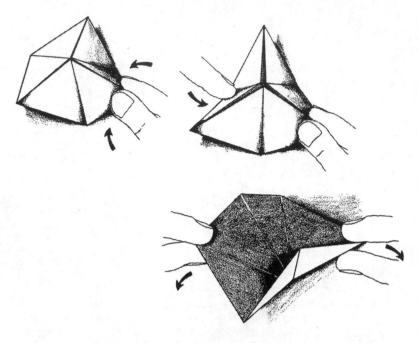

Figure 3. Trihexaflexagon is flexed by pinching together two of its triangles (*top*). The inner edge of the two opposite triangles may be opened with the other hand (*bottom*). If the figure cannot be opened, the adjacent pair of triangles is pinched. If the figure opens, it can be turned inside out, revealing a side that was not visible before. (Artist: Bunji Tagawa)

Now you fold the strip so that the same underside numbers face each other – 4 on 4, 5 on 5, 6 on 6, and so on. The resulting folded strip, illustrated in Figure 2B, is then folded back on the lines *ab* and *cd* (Figure 2C), forming the hexagon (Figure 2D); finally the blank triangle is turned under and pasted to the corresponding blank triangle on the other side of the strip. All this is easier to carry out with a marked strip of paper than it is to describe.

If you have made the folds properly, the triangles on one visible face of the hexagon will all be numbered 1, and on the other face they will all be numbered 2. Your hexahexaflexagon is now ready for flexing. Pinch two adjacent triangles together (see Figure 3), bending the paper along the line between them, and push in the opposite corner; the figure may then open up to face 3 or 5. By random flexing you should be able to find the other faces without much difficulty. Faces 4, 5, and 6 are a bit harder to uncover than faces 1, 2, and 3. At

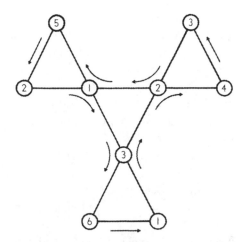

Figure 4. Diagram of a Tuckerman traverse on a hexahexaflexagon.

times you may find yourself trapped in an annoying cycle that keeps returning the same three faces over and over again.

Tuckerman quickly discovered that the simplest way to bring out all the faces of any flexagon was to keep flexing it at the same corner until it refused to open, then to shift to an adjacent corner. This procedure, known as the Tuckerman traverse, will bring up the six faces of a hexahexa in a cycle of 12 flexes, but faces 1, 2, and 3 turn up three times as often as faces 4, 5, and 6. A convenient way to diagram a Tuckerman traverse is shown in Figure 4, the arrows indicating the order in which the faces are brought into view. This type of diagram can be applied usefully to the traversing of any type of flexagon. When the model is turned over, a Tuckerman traverse runs the same cycle in reverse order.

By lengthening the chain of triangles, the committee discovered, one can make flexagons with 9, 12, 15 or more faces: Tuckerman managed to make a workable model with 48! He also found that with a strip of paper cut in a zigzag pattern (i.e., a strip with saw-tooth rather than straight edges) it is possible to produce a tetrahex-aflexagon (four faces) or a pentahexaflexagon. There are three different hexahexaflexagons – one folded from a straight strip, one from a chain bent into a hexagon, and one from a form that somewhat resembles a three-leaf clover. The decahexaflexagon (10 faces) has 82 different variations, all folded from weirdly bent strips. Flexagons can be formed with any desired number of faces, but beyond 10 the number of different species for each increases at an alarming rate. All even-numbered flexagons, by the way, are made of strips with two distinct sides, but those with an odd number of faces have only a single side, like a Moebius surface.

A complete mathematical theory of flexigation was worked out in 1940 by Tukey and Feynman. It shows, among other things, exactly how to construct a flexagon of any desired size or species. The theory has never been published, though portions of it have since been rediscovered by other mathematicians. Among the flexigators is Tuckerman's father, the distinguished physicist Louis B. Tuckerman, who was formerly with the National Bureau of Standards. Tuckerman senior devised a simple but efficient tree diagram for the theory.

The Japanese attack on Pearl Harbor called a halt to the committee's flexigation program, and war work soon scattered the four charter members to the winds. Stone became a lecturer in mathematics at the University of Manchester in England and is now at the University of Rochester. Feynman was a famous theoretical physicist at the California Institute of Technology. Tukey, a professor of mathematics at Princeton, has made brilliant contributions to topology and to statistical theory, which have brought him worldwide recognition. Tuckerman is a mathematician at IBM's research center in Yorktown Heights, New York.

One of these days, the committee hopes to reconvene and write a paper or two, which will be the definitive exposition of flexagon theory. Until then, the rest of us are free to flex our flexagons and see how much of the theory we can discover for ourselves.

ADDENDUM

In constructing flexagons from paper strips, it is a good plan to crease all the fold lines back and forth before folding the model. As a result, the flexagon flexes much more efficiently. Some readers have made more durable models by cutting triangles from poster board or metal and joining them with small pieces of tape or gluing them to one long piece of tape, leaving spaces between them for flexing. Louis Tuckerman keeps on hand a steel strip of such size that by

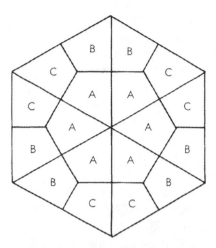

Figure 5. (Artist: Bunji Tagawa.)

wrapping paper tape of a certain width around it he can quickly produce a folded strip of the type shown in Figure 2. This saves considerable time in making flexagons from straight chains of triangles.

Readers have sent to me a large variety of ways in which flexagon faces could be decorated to make interesting puzzles or display striking visual effects. Each face of the hexahexa, for example, appears in at least two different forms, owing to a rotation of the component triangles relative to each other. Thus if we divide each face as shown in Figure 5, using different colors for the A, B, and C sections, the same face may appear with the A sections in the center as shown, or with the B or C sections in the center. Figure 6 shows how a geometrical pattern may be drawn on one face so as to appear in three different configurations.

Figure 6. (Artist: Bunji Tagawa.)

Of the 18 possible faces that can result from a rotation of the triangles, three are impossible to achieve with a hexahexa of the type made from a straight strip. This suggested to one correspondent the plan of pasting parts of three different pictures on each face so that by flexing the model properly, each picture could presumably be brought together at the center while the other two would be fragmented around the rim. On the three inner hexagons that cannot be brought together, he pasted the parts of three pictures of comely, undraped young ladies to make what he called a hexahexafrustragon. Another reader wrote that he achieved similar results by pasting together two adjacent triangular faces. This prevents one entire face from flexing into view, although the victim can see that it exists by peeking into the model's interior.

The statement that only 15 different patterns are possible on the straight-strip hexahexa must be qualified. An unsymmetrical coloring of the faces discloses the curious fact that three of these fifteen patterns have mirror-image partners. If you number the inner corners of each pattern with digits from 1 to 6, writing them in clockwise order, you will find that three of these patterns turn up with the same digits in counterclockwise order. Bearing this asymmetry in mind, one can say that the six faces of this hexahexa exhibit a total of 18 different configurations. This was first called to my attention by Albert Nicholas, professor of education at Monmouth College, Monmouth, Illinois, where the making of flexagons became something of a craze in the early months of 1957.

I do not know who first used a printed flexagon as an advertising premium or greeting card. The earliest sent to me was a trihexa distributed by the Rust Engineering Company of Pittsburgh to advertise their service award banquet in 1955. A handsome hexahexa, designed to display a variety of colored snowflake patterns, was used by *Scientific American* for its 1956 Christmas card.

For readers who would like to construct and analyze flexagons other than the two described in this chapter, here is a quick rundown on some low-order varieties.

1.   *The unahexa.* A strip of three triangles can be folded flat and the opposite ends joined to make a Moebius strip with a triangular edge. (For a more elegant model of a Moebius band

with triangular edge, see Chapter 7.) Since it has one side only, made up of six triangles, one might call it a unahexaflexagon, though of course it isn't six-sided and it doesn't flex.

2. *The duahexa.* Simply a hexagon cut from a sheet of paper. It has two faces but doesn't flex.

3. *The trihexa.* This has only the one form described in this chapter.

4. *The tetrahexa.* This flexagon likewise has only one form. It is folded from the crooked strip shown in Figure 7A.

5. *The pentahexa.* One form only. Folded from the strip in Figure 7B.

6. *The hexahexa.* There are three varieties, each with unique properties. One of them is described in this chapter. The other two are folded from the strips pictured in Figure 7C.

7. *The heptahexa.* This flexagon can be folded from the three strips shown in Figure 7D. The first strip can be folded in two different ways, making four varieties in all. The third form, folded from the overlapping figure-8 strip, is the first of what Louis Tuckerman calls the "street flexagons." Its faces can be numbered so that a Tuckerman traverse will bring uppermost the seven faces in serial order, like passing the street numbers on a row of houses.

The octahexa has 12 distinct varieties; the enneahexa has 27; and the decahexa, 82. The exact number of varieties of each order can be figured in more than one way, depending on how you define a distinct variety. For example, all flexagons have an asymmetric structure that can be right-handed or left-handed, but mirror-image forms should hardly be classified as different varieties. For details on the number of nonequivalent hexaflexagons of each order, consult the paper by Oakley and Wisner listed in the bibliography.

Straight chains of triangles produce only hexaflexagons with orders that are multiples of three. One variety of a 12-faced hexa is particularly easy to fold. Start with a straight chain twice as long as the one used for the hexahexa. "Roll" it into the form shown in Figure 2B. The strip is now the same length as the one used for the hexahexa. Fold this rolled strip exactly as if you were making a hexahexa. The result is a dodecahexaflexagon.

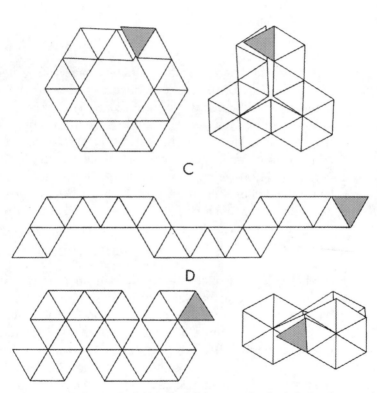

Figure 7. Crooked strips for folding hexaflexagons. The shaded triangles are tabs for pasting. (Artist: Bunji Tagawa)

In experimenting with higher-order flexagons, a handy rule to bear in mind is that the sum of the number of leaves (thicknesses of paper) in two adjacent triangular sections always equals the number of faces. It is interesting to note also that if each face of a flexagon is given a number or symbol, and the symbol marked on each triangular component, the order of symbols on the unfolded strip always exhibits a threefold symmetry. For example, the strip for the hexahexa in Figure 2 bears the following top and bottom pattern of digits:

$$123123 \quad 123123 \quad 123123$$
$$445566 \quad 445566 \quad 445566$$

A triple division similar to this is characteristic of all hexahexaflexagons, although one of the three divisions is always inverted on models of odd order.

Of the hundreds of letters I have received about flexagons, the following two were the most amusing. They appeared in the March and May issues of *Scientific American* in 1957.

Sirs:

I was quite taken with the article entitled "Flexagons" in your December issue. It took us only six or seven hours to paste the hexahexaflexagons together in the proper configuration. Since then it has been a source of continuing wonder.

But we have a problem. This morning one of our fellows was sitting flexing the hexahexaflexagon idly when the tip of his necktie became caught in one of the folds. With each successive flex, more of his tie vanished into the flexagon. With the sixth flexing he disappeared entirely.

We have been flexing the thing madly, and can find no trace of him, but we have located a sixteenth configuration of the hexahexaflexagon.

Here is our question: Does his widow draw workmen's compensation for the duration of his absence, or can we have him declared legally dead immediately? We await your advice.

Neil Uptegrove
Allen B. Du Mont Laboratories, Inc.
Clifton, N.J.

Sirs:

The letter in the March issue of your magazine complaining of the disappearance of a fellow from the Allen B. Du Mont Laboratories "down" a hexahexaflexagon, has solved a mystery for us.

One day, while idly flexing our latest hexahexaflexagon, we were confounded to find that it was producing a strip of multicolored material. Further flexing of the hexahexaflexagon finally disgorged a gum-chewing stranger. Unfortunately he was in a weak state and, owing to an apparent loss of memory, unable to give any account of how he came to be with us. His health has now been restored on our national diet of porridge, haggis, and whisky, and he has become quite a pet around the department, answering to the names of Eccles.

Our problem is, should we now return him and, if so, by what method? Unfortunately Eccles now cringes at the very sight of a hexahexaflexagon and absolutely refuses to "flex."

Robert M. Hill
The Royal College of Science and Technology
Glasgow, Scotland

POSTSCRIPT

The first book published on flexagons was *The Mysterious Flexagons* by Madeline Jones. This small book was written for children and contains nothing new on the topic.

Another book on flexagons is *Flexagons Inside Out* by Les (Leslie) Pook. Dr. Pook is a retired professor of mechanical engineering at University College London and the author of four books on engineering and more than a hundred technical papers. His book is a marvelous survey of flexagon theory, covering a wide variety of flexagons, including higher-dimensional forms that he calls "flexahedra."

I saw my first flexagon in the Manhattan apartment of Royal V. Heath, author of *Mathemagic* (a book of number tricks) and an amateur magician. He showed me a large colored cloth hexahexaflexagon and told me it had been invented by a group of graduate students in mathematics at Princeton University. I had recently sold *Scientific American* an article on logic machines and diagrams (which I later expanded into a book). It occurred to me that the magazine might like a piece on flexagons. I made a trip to Princeton, where I interviewed three members of the flexagon group: John Tukey, Louis Tuckerman, and his son Bryant. Stone had returned to England, and Feynman was in California. My article on flexagons ran in the December 1956 issue of *Scientific American*.

Gerard Piel, the magazine's publisher, called me to his office to ask if there was enough similar material on recreational math to justify a monthly column. I said there was. The column began in the January 1957 issue under the title "Mathematical Games" – a title given to it by the magazine's editors. By coincidence, its initials were the same as mine. The rest is history. My December article introduced flexagons to the general public, and all over Manhattan, especially in advertising offices, persons were happily folding and playing with these magical paper structures.

Since then, flexagon theory has made enormous progress. I later wrote a column on tetraflexagons, made with square "parts" instead of equilateral triangles. It is reprinted in *The Second "Scientific American" Book of Mathematical Games and Diversions.*

Today, surprising advances in flexagon theory have been made, and all sorts of fantastic varieties of flexagons have been and are still being invented. At the time I write, more than 50 Web sites on the Internet are devoted to the topic. I wouldn't be surprised if any day now someone forms a flexagon society that issues a periodical and holds an annual convention!

I've been told that the famous mathematician John Conway, at Princeton University, has made some breakthroughs in flexagon theory based on what he calls "stack permutations," but so far none of it has been published.

I despair of any effort to summarize the furious Internet activity. At the center of it all is Ann Schwartz, whose sisters Ela and Martha have a Web site that reports on Ann's constructions and other flexagon news. See Ela's Web site article "Flexagon Fever." There are even Web sites that sell flexagons and related structures. A magic supply firm called No Way is selling for $95 a 19-inch diameter hexahexaflexagon made of styrene. Its faces have six different colors. With it come sample stage routines including one based on a gospel story.

BIBLIOGRAPHY

"Hexahexaflexagrams." Margaret Joseph in *Mathematics Teacher* 44 (April 1951): 247–48. Tells how to make a straight-chain hexahexaflexagon.

"A Six-Sided Hexagon." William R. Ransom in *School Science and Mathematics* 52 (1952): 94. Tells how to make a trihexaflexagon.

"The Flexagon and the Hexahexaflexagram." F. G. Maunsell in *Mathematical Gazette* 38 (1954): 213–214. Describes the trihexa and hexahexa. See also vol. 41 (1957): 55–56, for a note on this article by Joan Crampin. It describes hexaflexagons folded from straight strips (orders 3, 6, 9, 12, . . .).

"Flexagons." Cletas Oakley and Robert Wisner in *American Mathematical Monthly* 64 (March 1957): 143–254.

"The Flexagon Family." Roger F. Wheeler in *Mathematical Gazette* 42 (February 1958): 1–6. A fairly complete analysis of straight- and crooked-strip forms.

"How to Construct Hexaflexagons." Sidney Scott in *Recreational Mathematics Magazine* 12 (December 1962): 43–49.

"Protean Shapes with Flexagons." William Ransom in *Recreational Mathematics Magazine* 13 (February 1963): 35–37.

"The Construction of Flexagons." Pamela Liebeck in *Mathematical Gazette* 48 (December 1964): 397–402.

The Mysterious Flexagons. Madeline Jones. Crown, 1966. A book for children printed on heavy paper so the child can cut strips for folding.

Mathematics on Vacation. Joseph Madachy. Scribner's, 1966, 62–84.

"Hybrid Flexagons." Douglas Engel in *Journal of Recreational Mathematics* 2 (January 1969): 35–41.

"Sneaking Up On a Group." Jean Pederson in *The Two-Year College Mathematics Journal* 3 (1972): 9–12.

Mathemagic with Flexagons. Donovan Johnson. Activity Resources, 1974.

"Classifying and Counting Hexaflexagrams." Thomas O'Reilly in *Journal of Recreational Mathematics* 8 (1975): 182–187.

"Symmetries of the Trihexaflexagon." Michael Gilpin in *Mathematics Magazine* 49 (September 1976): 189–192.

Flexagons. Paul Jackson. British Origami Society, 1978.

"V-Flexing the Hexahexaflexagon." T. Bruce McLean in *American Mathematical Monthly* 86 (June–July 1979): 457–466.

"Hexplay." Scot Morris in *Omni* (October 1984): 89ff.

"To Flexland With Mr. Flexman." A. Panov in *Quantum* (March–April 1992): 64–65.

"Flexland Revisited." A. Panov and A. Keltnin in *Quantum* (July–August 1993): 64–65.

"Hexaflexagons + HHPG = Slipagon." Douglas Engel in *Journal of Recreational Mathematics* 3 (1995): 161–166.

"Can You Do It With Heptagons?" A. Hirst in *Mathematical Gazette* 79 (1995): 17–29.

"Note on Straight Strip Flexagons." J. Crampin in *Mathematical Gazette* 41 (1997): 55–56.

"The Faces of the Trihexaflexagon." Peter Hilton, Jean Pederson, and Hans Walser in *Mathematics Magazine* 70 (October 1997): 243–251.

Build Your Own Polyhedra. Peter Hilton and Jean Pederson. Addison-Wesley, 1998.

The Magic of Flexagons. David Mitchell. Tarquin, 1998.

Flexagons Inside Out. Les Pook. Cambridge University Press, 2003.

"Loops of Regular 2^n-gons." K. R. McLean and J. R. Sylvester in *Mathematical Gazette* 87 (2003): 512–513.

"The Nonahexaflexagon." T. Bruce McLean and Emily E. McLean in *Journal of Recreational Mathematics* 32 (2003–2004): 56–61.

"The Hexa-Dodeca-Flexagon." Ann Schwartz and Jeff Busby in *G4G7 Gathering 4 Gardner,* Vol. 1 (2007): 93–96.

Elements of Flexagon Theory. Frank Bernhart. Unpublished monograph.

The Curious World of Flexagons. Phil Mitchell. Unpublished book.

CHAPTER TWO

Magic with a Matrix

MAGIC SQUARES HAVE INTRIGUED MATHEMATICIANS for more than 2,000 years. In the traditional form, the square is constructed so that the numbers in each row, each column, and each diagonal add up to the same total. However a magic square of an entirely different type is pictured in Figure 8. This square seems to have no system: The numbers appear to be distributed in the matrix at random. Nevertheless, the square possesses a magical property as astonishing to most mathematicians as it is to laymen.

A convenient way to demonstrate this property is to equip yourself with five pennies and 20 little paper markers (say pieces of paper matches). Now ask someone to pick any number in the square. Lay a penny on this number and eliminate all the other numbers in the same row and in the same column by covering them with markers.

Ask your spectator to pick a second number by pointing to any uncovered cell. As before, put a penny on this number and cover all the others in the same row and column. Repeat this procedure twice more. One uncovered cell will remain. Cover it with the fifth penny.

When you add the five numbers beneath the pennies – numbers chosen seemingly at random – the total is certain to be 57. This is no accident. The total will be the same with every repetition of the experiment.

If you enjoy solving mathematical puzzles, you may wish to pause at this point to analyze the square and see if you can discover its secret yourself.

Like most tricks, this one is absurdly simple when explained. The square is nothing more than an old-fashioned addition table,

16

19	8	11	25	7
12	1	4	18	0
16	5	8	22	4
21	10	13	27	9
14	3	6	20	2

Figure 8

arranged in a tricky way. The table is generated by two sets of numbers: 12, 1, 4, 18, 0 and 7, 0, 4, 9, 2. The sum of these numbers is 57. If you write the first set of numbers horizontally above the top row of the square, and the second set vertically beside the first column (see Figure 9), you can see at once how the numbers in the cells are determined. The number in the first cell (top row, first column) is the sum of 12 and 7, and so on through the square.

You can construct a magic square of this kind as large as you like and with any combination of numbers you choose. It does not matter in the least how many cells the square contains or what

	12	1	4	18	0
7	19	8	11	25	7
0	12	1	4	18	0
4	16	5	8	22	4
9	21	10	13	27	9
2	14	3	6	20	2

Figure 9

	1	2	3	4
0	1	2	3	4
4	5	6	7	8
8	9	10	11	12
12	13	14	15	16

Figure 10

numbers are used for generating it. They may be positive or negative, integers or fractions, rationals or irrationals. The resulting table will always possess the magic property of forcing a number by the procedure described, and this number will always be the sum of the two sets of numbers that generate the table. In the case given here, you could break the number 57 into any eight numbers that add up to that sum.

The underlying principle of the trick is now easy to see. Each number in the square represents the sum of a pair of numbers in the two generating sets. That particular pair is eliminated when a penny is placed on the number. The procedure forces each penny to lie in a different row and column. Thus the five pennies cover the sums of five different pairs of the ten generating numbers, which is the same as the sum of all ten numbers.

One of the simplest ways to form an addition table on a square matrix is to start with 1 in the upper left corner and then continue from left to right with integers in serial order. A 4×4 matrix of this sort becomes an addition table for the two sets of numbers 1, 2, 3, 4, and 0, 4, 8, 12 (Figure 10). This matrix will force the number 34.

The forced number is of course a function of the size of the square. If n is the number of cells on a side, then the forced number will be

$$\frac{n^3 + n}{2}$$

On squares with an odd number of cells on the side, this forced number will equal the product of n and the number on the center cell.

If you start with a number higher than 1 (call it a) and continue in serial order, the forced number will be

$$\frac{n^3 + n}{2} + n(a - 1)$$

It is interesting to note that the forced number is the same as the total of each row and column on a traditional magic square that is formed from the same numerical elements.

By means of the second formula, it is easy to calculate the starting number for a matrix of any desired size that will force any desired number. An impressive impromptu stunt is to ask someone to give you a number above 30 (this is specified to avoid bothersome minus numbers in the matrix), then proceed to draw quickly a 4 × 4 matrix that will force that number. (Instead of using pennies and paper markers, a faster procedure is to let the spectator circle each chosen number and then draw a line through its row and column.)

The only calculation you need to make (it can be done in your head) is to subtract 30 from the number he names and then divide by 4. For example, he calls out 43. Subtracting 30 gives 13. Dividing 13 by 4 results in $3^1/_4$. If you put this number in the first cell of a 4 × 4 matrix and continue in serial order with $4^1/_4$, $5^1/_4$, ..., you will produce a magic square that will force 43.

To make the square more baffling, however, the order of the numbers should be scrambled. For instance, you might put the first number, $3^1/_4$ in a cell in the third row as shown in Figure 11, and the next three numbers ($4^1/_4$, $5^1/_4$, and $6^1/_4$) in the same row in a random order. Now you may write the next four numbers in another row (it does not matter which), but they must be in the same cell sequence you followed before. Do exactly the same with the last two rows. The final result will be something like the square shown in Figure 12.

If you want to avoid fractions and still force the number 43, you can drop the $^1/_4$ after all of the numbers and add 1 to each of the four highest whole numbers, making them 16, 17, 18, and 19. Similarly you would add 2 to these numbers if the fraction were $^2/_4$, or 3 if it were $^3/_4$.

Figure 11

Interchanging the order in rows or in columns has no effect on the square's magic property, and by scrambling the cells in this manner, you make the matrix appear much more mysterious than it really is.

Multiplication tables may also be used to force a number. In this case, the chosen numbers must be multiplied instead of added. The final product will equal the product of the numbers used to generate the table.

I have not been able to discover who first applied this delightful property of addition and multiplication tables to a trick. A parlor stunt with numbered cards, based on the principle, was published by Maurice Kraitchik on page 184 of his *Mathematical Recreations*, 1942. This is the earliest reference I have found to the principle. Since 1942, several mathematically inclined conjurers have

$16\frac{1}{4}$	$18\frac{1}{4}$	$15\frac{1}{4}$	$17\frac{1}{4}$
$8\frac{1}{4}$	$10\frac{1}{4}$	$7\frac{1}{4}$	$9\frac{1}{4}$
$4\frac{1}{4}$	$6\frac{1}{4}$	$3\frac{1}{4}$	$5\frac{1}{4}$
$12\frac{1}{4}$	$14\frac{1}{4}$	$11\frac{1}{4}$	$13\frac{1}{4}$

Figure 12

introduced variations on the theme. For instance, Mel Stover of Winnipeg observed that if you draw a square around 16 numbers on any calendar page, the square forms an addition table that forces a number twice the sum of the two numbers at either of the diagonally opposite corners.

The use of playing cards also opens up colorful possibilities. For example, is it possible to arrange a deck so that it can be cut and a square array of cards dealt from the cut that will always force the same number? The principle is relatively unexplored and may have many curious ramifications yet to be discovered.

ADDENDUM

Stewart James, a magician in Courtright, Ontario, devised a novel variation of the magic square in which one can force any desired word on an audience. Suppose you wish to force the word JAMES. You form a square of 25 cards, the undersides of which (unknown to anyone but you) bear letters as follows:

```
J  A  M  E  S
J  A  M  E  S
J  A  M  E  S
J  A  M  E  S
J  A  M  E  S
```

Someone is asked to pick one of the cards by touching its back. This card is placed aside, without showing its face, and all other cards in the same row and column are removed. This procedure is repeated three more times, then the one remaining card is placed with the other four that have been selected. The five cards are then turned over and arranged to spell JAMES. The procedure makes it impossible, of course, for the five selected cards to include duplicates.

One reader wrote that he found the magic square an intriguing curiosity to draw on birthday cards for mathematically minded friends. The recipient follows instructions, adds his chosen numbers, and is startled to find that the total is his age.

Workers

Machines	A 12	B 1	C 4	D 18	E 0
arcer 7	19	8	11	25	7
boinker 0	12	1	4	18	0
clanker 4	16	5	8	22	4
denter 9	21	10	13	27	9
edger 2	14	3	6	20	2

Figure 13. A simple assignment problem.

AFTERWORD, 1988

Philip Goldstein, a magician who performs under the stage name of Max Maven, thought of a clever way to present the magic square trick that avoids forcing the last number. He gives a spectator felt-tip pens of n colors where n is the square's order. He then draws a horizontal line through each row, using a different color for each row and permuting the colors any way he likes. He then does the same thing for each column. The numbers in the cells where horizontal and vertical lines of the same color intersect are added. The sum is, of course, the predetermined number.

POSTSCRIPT

Peter Renz called my attention to the close relationship between magic matrices and what in the field of operations research are called assignment problems.

Consider the 5×5 matrix shown in Figure 13. Along the top are initials of the names of five workers – Ann, Ben, Charles, Doris, and

Emily – at a plant that makes widgets. Under each initial a number represents the worker's productivity during, say, an hour. On the left side of the matrix are the names of five machines. The accompanying numbers represent the machine's productivity in an hour.

Each number within the matrix stands for the combined productivity when a worker is assigned to a particular machine. This number is the sum of the machine's number at the left of the row, and the worker's number at the top of the column.

The matrix clearly is magic! In this case, it forces the number 57 after five numbers are circled and added. Because 57 is a constant, no matter how workers are assigned to machines their total productivity is certain to be 57.

Two good discussions of more advanced assignment problems can be found in George B. Danzig, *Linear Programming and Extensions* (Princeton University Press, 1963), and L. R. Ford, Jr., and D. R. Fulkerson, *Flows in Networks* (Princeton University Press, 1962).

BIBLIOGRAPHY

"Webster Had a Word for It." Stewart James in *Linking Ring* (an American magic magazine), October 1952.

"And So Force." P. Howard Lyons in *Genii* 19 (an American magic magazine), February 1955: 19.

"OGNIB." Mel Stover in *Ibidem* 7 (a Canadian magic magazine), September 1956: 7.

"The Irresistible Force." Mel Stover in *New Phoenix* 340 (an American magic magazine), January 1957: 340.

"Rainbow Matrix." Philip Goldstein in *The Violet Book of Mentalism*. Privately published for the magic trade, 1980.

"Force Majeur." Philip Goldstein in *Doth*. Privately published, 1987. Explains a simple way of drawing in your head a matrix that will force a number called out by a spectator.

CHAPTER THREE

Nine Problems

1. THE RETURNING EXPLORER

An old riddle runs as follows. An explorer walks one mile due south, turns and walks one mile due east, turns again and walks one mile due north. He finds himself back where he started. He shoots a bear. What color is the bear? The time-honored answer is "White" because the explorer must have started at the North Pole. Not long ago, however, someone made the discovery that the North Pole is not the only starting point that satisfies the given conditions! Can you think of any other spot on the globe from which one could walk a mile south, a mile east, a mile north and find himself back at his original location?

2. DRAW POKER

Two men play a game of draw poker in the following curious manner. They spread a deck of 52 cards face up on the table so that they can see all the cards. The first player draws a hand by picking any five cards he chooses. The second player does the same. The first player now may keep his original hand or draw up to five cards. His discards are put aside out of the game. The second player may now draw likewise. The person with the higher hand then wins. Suits have equal value, so that two flushes tie unless one is made of higher cards. After a while, the players discover that the first player can always win if he draws his first hand correctly. What hand must this be?

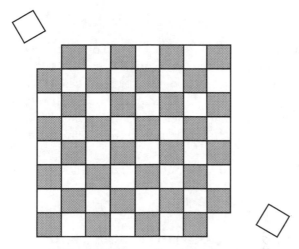

Figure 14. The mutilated chessboard.

3. THE MUTILATED CHESSBOARD

The props for this problem are a chessboard and 32 dominoes. Each domino is of such size that it exactly covers two adjacent squares on the board. The 32 dominoes therefore can cover all 64 of the chessboard squares. But now suppose we cut off two squares at diagonally opposite corners of the board (see Figure 14) and discard one of the dominoes. Is it possible to place the 31 dominoes on the board so that all the remaining 62 squares are covered? If so, show how it can be done. If not, prove that it is impossible.

4. THE FORK IN THE ROAD

Here's a recent twist on an old type of logic puzzle. A logician vacationing in the South Seas finds himself on an island inhabited by the two proverbial tribes of liars and truth-tellers. Members of one tribe always tell the truth, members of the other always lie. He comes to a fork in a road and has to ask a native bystander which branch he should take to reach a village. He has no way of telling whether the native is a truth-teller or a liar. The logician thinks a moment and then asks *one* question only. From the reply, he knows which road to take. What question does he ask?

5. SCRAMBLED BOX TOPS

Imagine that you have three boxes, one containing two black marbles, one containing two white marbles, and the third, one black marble and one white marble. The boxes were labeled for their contents – BB, WW, and BW – but someone has switched the labels so that every box is now incorrectly labeled. You are allowed to take one marble at a time out of any box, without looking inside, and by this process of sampling you are to determine the contents of all three boxes. What is the smallest number of drawings needed to do this?

6. BRONX VS. BROOKLYN

A young man lives in Manhattan near a subway express station. He has two girl friends, one in Brooklyn and one in the Bronx. To visit the girl in Brooklyn, he takes a train on the downtown side of the platform; to visit the girl in the Bronx, he takes a train on the uptown side of the same platform. He likes both girls equally well, so he simply takes the first train that comes along. In this way he lets chance determine whether he rides to the Bronx or to Brooklyn. The young man reaches the subway platform at a random moment each Saturday afternoon. Brooklyn and Bronx trains arrive at the station equally often – every 10 minutes. Yet for some obscure reason, he finds himself spending most of his time with the girl in Brooklyn:

In fact on the average, he goes there nine times out of ten. Can you think of a good reason why the odds so heavily favor Brooklyn ?

7. CUTTING THE CUBE

A carpenter, working with a buzz saw, wishes to cut a wooden cube, three inches on a side, into 27 one-inch cubes. He can do this easily by making six cuts through the cube, keeping the pieces together in the cube shape (see Figure 15). Can he reduce the number of necessary cuts by rearranging the pieces after each cut?

8. THE EARLY COMMUTER

A commuter is in the habit of arriving at his suburban station each evening exactly at five o'clock. His wife always meets the train and

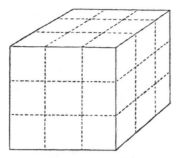

Figure 15. The sliced cube.

drives him home. One day he takes an earlier train, arriving at the station at four o'clock. The weather is pleasant, so instead of telephoning home, he starts walking along the route always taken by his wife. They meet somewhere on the way. He gets into the car, and they drive home, arriving at their house 10 minutes earlier than usual. Assuming that the wife always drives at a constant speed, and that on this occasion she left just in time to meet the five o'clock train, how long did the husband walk before his wife picked him up?

9. THE COUNTERFEIT COINS

In recent years, a number of clever coin-weighing or ball-weighing problems have aroused widespread interest. Here is a new and charmingly simple variation. You have 10 stacks of coins, each consisting of 10 half-dollars (see Figure 16). One entire stack is counterfeit, but you do not know which one. You do know the weight of a genuine half-dollar, and you are also told that each counterfeit coin weighs one gram more than it should. You may weigh the coins on a pointer scale. What is the smallest number of weighings necessary to determine which stack is counterfeit?

ANSWERS

1. Is there any other point on the globe, besides the North Pole, from which you could walk a mile south, a mile east, and a mile north and find yourself back at the starting point? Yes indeed, there is not just one point but an infinite number of them! You could start from any point on a circle drawn around the South Pole at a distance slightly more than $1 + \frac{1}{2\pi}$ miles (about

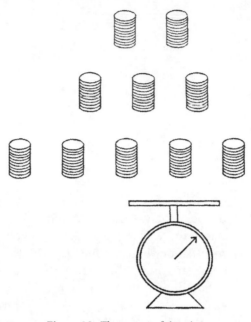

Figure 16. The counterfeit coins.

1.16 miles) from the Pole – the distance is "slightly more" to take into account the curvature of the earth. After walking a mile south, your next walk of one mile east will take you on a complete circle around the Pole, and the walk one mile north from there will then return you to the starting point. Thus your starting point could be any one of the infinite number of points on the circle with a radius of about 1.16 miles from the South Pole. But this is not all. You could also start at points closer to the Pole, so that the walk east would carry you just twice around the Pole, or three times, and so on.

Suppose we ignore the restrictions that the explorer walks south, east, and then north. He walks a mile in any direction, turns 90 degrees, goes another mile, turns 90 degrees, walks a mile and finds himself back where he started. Where does he start? The answer, of course, is that he can start anywhere.

2.   There are 88 winning first hands. They fall into two categories: (1) four tens and any other card (48 hands); (2) three tens and any of the following pairs from the suit not represented by a ten:

A-9, K-9, Q-9, J-9, K-8, Q-8, J-8, Q-7, J-7, J-6 (40 hands). The second category was called to my attention by two readers: Charles C. Foster of Princeton, New Jersey, and Christine A. Peipers of New York. I have never seen these hands included in any previously published answers to the problem.

3. It is impossible to cover the mutilated chessboard (with two opposite corner squares cut off) with 31 dominoes, and the proof is easy. The two diagonally opposite corners are of the same color. Therefore their removal leaves a board with two more squares of one color than of the other. Each domino covers two squares of opposite color, since only opposite colors are adjacent. After you have covered 60 squares with 30 dominoes, you are left with two uncovered squares of the same color. These two cannot be adjacent; therefore, they cannot be covered by the last domino.

Ralph Gomory asked himself, if two cells of *opposite* color are removed from anywhere on a chessboard, can the remaining area be covered by 31 dominoes? His beautiful proof that the answer is "yes" can be found in Chapter 16 of my *Unexpected Hanging*.

4. If we require that the question be answerable by "yes" or "no," there are several solutions, all exploiting the same basic gimmick. For example, the logician points to one of the roads and says to the native, "If I were to ask you if this road leads to the village, would you say 'yes'?" The native is forced to give the right answer, even if he is a liar! If the road does lead to the village, the liar would say "no" to the direct question, but as the question is put, he lies and says he would respond "yes." Thus the logician can be certain that the road does lead to the village, whether the respondent is a truth-teller or a liar. On the other hand, if the road actually does not go to the village, the liar is forced in the same way to reply "no" to the inquirer's question.

A similar question would be, "If I asked a member of the other tribe whether this road leads to the village, would he say 'yes'?" To avoid the cloudiness that results from a question within a question, perhaps this phrasing (suggested by Warren C. Haggstrom, of Ann Arbor, Michigan) is best: "Of the two statements,

'You are a liar' and 'This road leads to the village,' is one and only one of them true?" Here again, a "yes" answer indicates it is the road, a "no" answer that it isn't, regardless of whether the native lies or tells the truth.

Dennis Sciama, Cambridge University cosmologist, and John McCarthy of Hanover, New Hampshire, called my attention to a delightful additional twist on the problem. "Suppose," Mr. McCarthy wrote (in a letter published in *Scientific American*, April 1957), "the logician knows that 'pish' and 'tush' are the native words for 'yes' and 'no' but has forgotten which is which, though otherwise he can speak the native language. He can still determine which road leads to the village.

"He points to one of the roads and asks, 'If I asked you whether the road I am pointing to is the road to the village would you say pish?' If the native replies, 'Pish,' the logician can conclude that the road pointed to is the road to the village even though he will still be in the dark as to whether the native is a liar or a truth-teller and as to whether 'pish' means yes or no. If the native says, 'Tush,' he may draw the opposite conclusion."

H. Janzen of Queens University, Kingston, Ontario, and several other readers informed me that if the native's answer does not have to be "yes" or "no," there is a question that reveals the correct road regardless of how many roads meet at the intersection. The logician simply points to all the roads, including the one he has just traveled, and asks, "Which of these roads leads to the village?" The truth-teller points to the correct one, and the liar presumably points to all the others. The logician could also ask, "Which roads do not lead to the village?" In this case the liar would presumably point only to the correct one. Both cases, however, are somewhat suspect. In the first case the liar might point to only one incorrect road and in the second case he might point to several roads. These responses would be lies in a sense, though one would not be the strongest possible lie and the other would contain a bit of truth.

The question of how precisely to define "lying" even enters into the previous "yes" and "no" solutions. I know of no better way to make this clear than by quoting in full the following

letter, which *Scientific American* received from Willison Crich-
ton and Donald E. Lamphiear, both of Ann Arbor, Michigan.

It is a sad commentary on the rise of logic that it leads to the
decay of the art of lying. Even among liars, the life of reason
seems to be gaining ground over the better life. We refer to puzzle number 4 in the February issue, and its solution. If we accept
the proposed solution, we must believe that liars can always
be made the dupes of their own principles, a situation, indeed,
which is bound to arise whenever lying takes the form of slavish
adherence to arbitrary rules.

For the anthropologist to say to the native, "If I were to ask you
if this road leads to the village, would you say 'yes'?" expecting
him to interpret the question as counterfactual conditional in
meaning as well as form, presupposes a certain preciosity on the
part of the native. If the anthropologist asks the question casu-
ally, the native is almost certain to mistake the odd phraseology
for some civility of manner taught in Western democracies, and
answer as if the question were simply, "Does this road lead to
the village?" On the other hand, if he fixes him with a glitter-
ing eye in order to emphasize the logical intent of the question,
he also reveals its purpose, arousing the native's suspicion that
he is being tricked. The native, if he is worthy the name of liar,
will pursue a method of counter-trickery, leaving the anthropologist misinformed. On this latter view, the proposed solution is
inadequate, but even in terms of strictly formal lying, it is faulty
because of its ambiguity.

The investigation of unambiguous solutions leads us to a
more detailed analysis of the nature of lying. The traditional definition employed by logicians is that a liar is one who always says
what is false. The ambiguity of this definition appears when we
try to predict what a liar will answer to a compound truth func-
tional question, such as, "Is it true that if this is the way to town,
you are a liar?" Will he evaluate the two components correctly
in order to evaluate the function and reverse his evaluation in
the telling, or will he follow the impartial policy of lying to himself as well as to others, reversing the evaluation of each component before computing the value of the function, and then
reversing the computed value of the function? Here we distinguish the *simple liar* who always utters what is simply false from
the *honest liar* who always utters the logical dual of the truth.

The question, "Is it true that if this is the way to town, you are a liar?" is a solution if our liars are honest liars. The honest liar and the truth-teller both answer "yes" if the indicated road is not the way to town, and "no" if it is. The simple liar, however, will answer "no" regardless of where the village is. By substituting equivalence for implication we obtain a solution which works for both simple and honest liars. The question becomes, "Is it true that this is the way to town if and only if you are a liar?" The answer is uniformly "no" if it the way, and "yes" if it is not.

But no lying primitive savage could be expected to display the scrupulous consistency required by these conceptions, nor would any liar capable of such acumen be so easily outwitted. We must therefore consider the case of the *artistic liar* whose principle is always to deceive. Against such an opponent the anthropologist can only hope to maximize the probability of a favorable outcome. No logical question can be an infallible solution, for if the liar's principle is to deceive, he will counter with a strategy of deception which circumvents logic. Clearly the essential feature of the anthropologist's strategy must be its psychological soundness. Such a strategy is admissible since it is even more effective against the honest and the simple liar than against the more refractory artistic liar.

We therefore propose as the most general solution the following question or its moral equivalent, "Did you know that they are serving free beer in the village?" The truth-teller answers "no" and immediately sets off for the village, the anthropologist following. The simple or honest liar answers "yes" and sets off for the village. The artistic liar, making the polite assumption that the anthropologist is also devoted to trickery, chooses his strategy accordingly. Confronted with two contrary motives, he may pursue the chance of satisfying both of them by answering, "Ugh! I hate beer!" and starting for the village. This will not confuse a good anthropologist. But if the liar sees through the ruse, he will recognize the inadequacy of this response. He may then make the supreme sacrifice for the sake of art and start down the wrong road. He achieves a technical victory, but even so, the anthropologist may claim a moral victory, for the liar is punished by the gnawing suspicion that he has missed some free beer.

5.   You can learn the contents of all three boxes by drawing just one marble. The key to the solution is your knowledge that the

labels on all three of the boxes are incorrect. You must draw a marble from the box labeled BW. Assume that the marble drawn is black. You know then that the other marble in this box must be black also; otherwise, the label would be correct. Since you have now identified the box containing two black marbles, you can at once tell the contents of the box marked WW: You know it cannot contain two white marbles because its label has to be wrong; it cannot contain two black marbles, for you have identified that box; therefore, it must contain one black and one white marble. The third box, of course, must then be the one holding two white marbles. You can solve the puzzle by the same reasoning if the marble you draw from the BW box happens to be white instead of black.

6. The answer to this puzzle is a simple matter of train schedules. Even though the Brooklyn and Bronx trains arrive equally often – at 10-minute intervals – it happens that their schedules are such that the Bronx train always comes to this platform one minute after the Brooklyn train. Thus the Bronx train will be the first to arrive only if the young man happens to come to the subway platform during this one-minute interval. If he enters the station at any other time (i.e., during a nine-minute interval), the Brooklyn train will come first. Since the young man's arrival is random, the odds are 9:1 for Brooklyn.

7. There is no way to reduce the cuts to fewer than six. This is at once apparent when you focus on the fact that a cube has six sides. The saw cuts straight – one side at a time. To cut the one-inch cube at the center (the one that has no exposed surfaces to start with) must take six passes of the saw.

   This problem was originated by Frank Hawthorne, supervisor of mathematics education, State Department of Education, Albany, New York, and first published in *Mathematics Magazine*, September–October 1950 (Problem Q-12).

   Cubes of $2 \times 2 \times 2$ and $3 \times 3 \times 3$ are unique in the sense that regardless of how the pieces are rearranged before each cut (provided each piece is cut somewhere), the former will always require three cuts and the latter six to slice into unit cubes.

   The $4 \times 4 \times 4$ cube requires nine cuts if the pieces are kept together as a cube, but by proper piling before each cut, the number of cuts can be reduced to six. If at each piling you see that every piece is cut as nearly in half as possible, the minimum number of cuts will be achieved. In general, for an $n \times n \times n$ cube, the minimum number of cuts is $3k$ where k is defined by

$$2^k \geq n > 2^{k-1}$$

This general problem was posed by L. R. Ford, Jr., and D. R. Fulkerson, both of The Rand Corporation, in the *American Mathematical Monthly*, August–September 1957 (Problem E1279), and answered in the March 1958 issue. The problem is a special case of a more general problem (the minimum cuts for slicing an $a \times b \times c$ block into unit cubes) contributed by Leo Moser, of the University of Alberta, to *Mathematics Magazine*, Vol. 25, March–April 1952, p. 219.

Eugene J. Putzer and R. W. Lowen generalized the problem still further in a research memorandum, "On the Optimum Method of Cutting a Rectangular Box into Unit Cubes," issued in 1958 by Convair Scientific Research Laboratory, San Diego. The authors considered blocks of n-dimensions, with integral sides, which are to be sliced by a minimum number of planar cuts into unit hypercubes. In three dimensions, the problem is one that the authors feel might "have important applications in the cheese and sugar-loaf industries."

8. The commuter has walked for 55 minutes before his wife picks him up. Since they arrive home 10 minutes earlier than usual, this means that the wife has chopped 10 minutes from her usual travel time to and from the station, or 5 minutes from her travel time to the station. It follows that she met her husband 5 minutes before his usual pick-up time of five o'clock, or at 4:55. He started walking at four o'clock; therefore, he walked for 55 minutes. The man's speed of walking, the wife's speed of driving, and the distance between home and station are not needed for solving the problem. If you tried to solve it by juggling figures for these variables, you probably found the problem exasperating.

When this problem was presented in *Scientific American*, it was unfortunately worded, suggesting that the wife habitually arrived early at the station and waited for the five-o'clock train.

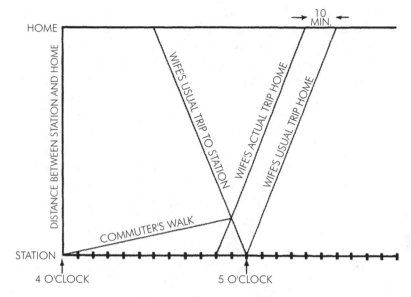

Figure 17. Graph of the commuter problem.

If this is the case, the husband's walking time lies within a range of 50 to 55 minutes.

A number of readers pointed out that the problem yields readily to solution by what Army logisticians call a march graph (see Figure 17). Time is plotted on the horizontal axis; distance on the vertical. The graph shows clearly that the wife could leave home up to 10 minutes earlier than the leaving time required to just meet the train. The lower limit (50 minutes) of her husband's walking time can occur only when the wife leaves a full 10 minutes earlier and either drives habitually at infinite speed (in which case her husband arrives home at the same moment she leaves), or the husband walks at an infinitesimal speed (in which case she meets him at the station after he has walked 50 minutes and gotten nowhere). "Neither image rings false," wrote David W. Weiser, assistant professor of natural science at the University of Chicago, in one of the clearest analyses I received of the problem, "considering the way of a wife with a car, or of a husband walking past a tavern."

9. The counterfeit stack can be identified by a single weighing of
    coins. You take one coin from the first stack, two from the sec-
    ond, three from the third, and so on until you have taken all ten
    coins of the tenth stack. You then weigh the whole sample col-
    lection on the pointer scale. The excess weight of this collection,
    in number of grams, corresponds to the number of the coun-
    terfeit stack. For example, if the group of coins weighs 7 grams
    more than it should, then the counterfeit stack must be the sev-
    enth one, from which you took seven coins (each weighing 1
    gram more than a genuine half-dollar). Even if there had been
    an eleventh stack of ten coins, the procedure just described
    would still work, for no excess weight would indicate that the
    one remaining stack was counterfeit.

CHAPTER FOUR

Ticktacktoe

WHO HAS NOT AS A CHILD PLAYED TICKTACKTOE, that most ancient and universal struggle of wits of which Wordsworth wrote (*Prelude*, Book I):

> At evening, when with pencil, and smooth slate
> In square divisions parcelled out and all
> With crosses and with cyphers scribbled o'er,
> We schemed and puzzled, head opposed to head
> In strife too humble to be named in verse.

At first sight it is not easy to understand the enduring appeal of a game that seems no more than child's play. While it is true that even in the simplest version of the game, the number of possible moves is very large – 15,120 ($9 \times 8 \times 7 \times 6 \times 5$) different sequences for the first five moves alone – there are really only a few basic patterns. Any astute youngster can become an unbeatable player by analyzing the game for an hour or so. But ticktacktoe also has its more complex variations and strategic aspects.

In the lingo of game theory, ticktacktoe is a two-person contest that is "finite" (comes to a definite end), has no element of chance, and is played with "perfect information," all moves being known to both players. If played "rationally" by both sides, the game must end in a draw. The only chance of winning is to catch an unwary opponent in a "trap" where a row can be scored on the next move in two ways, only one of which can be blocked.

Of the three possible opening plays – a corner, the center or a side box – the strongest opening is the corner because the opponent can avoid being trapped at the next move only by one of the eight

Figure 18. The first player (X) has a choice of three openings. To avoid losing, second player (O) must choose one of the cells indicated.

possible choices: the center. Conversely, center opening traps can be blocked only by seizing a corner. The side opening, in many ways the most interesting because of its richness in traps on both sides, must be met by taking one of four cells. The three openings and the possible responses by a second player who plays rationally are diagramed in Figure 18.

Variants of ticktacktoe that are mathematically more exciting than the present form were played many centuries before the Christian era. All of them employ six counters and can be played on the board pictured in Figure 19 – with one player using three pennies, the other, three dimes. In the simplest form, popular in ancient China, Greece, and Rome, players take turns placing a counter on the board until all six are down. If neither player has won by getting three in a row (orthogonally or diagonally) they continue playing by moving on each turn a single counter to any adjacent square. Only moves along the orthogonals are permitted.

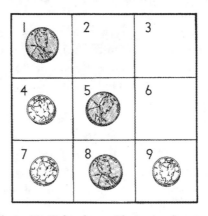

Figure 19. Ticktacktoe with moving counters.

Ovid mentions this game in Book III of his *Art of Love*, including it among a group of games that he advises women to master if they wish to be popular with men. The game was common in England in 1300 when it was called three men's morris, the ancestor of 9, 11, and 12 men's morris, or "mill" as it is usually called in the United States today. Because the first player has a sure win by playing first in the center, this opening is usually barred. With this restriction, the game is a draw if played rationally, but it swarms with potential traps on both sides.

A variation of this game permits moves to neighboring cells along the two main diagonals of the square. A further extension (attributed to early American Indians) allows any counter to move one step in any direction, orthogonally or diagonally (e.g., a move can be made from cell 2 to cell 4). In the first version the initial player can still force a win if allowed to open on the center, but the second variant is probably a draw. A free-wheeling version called *les pendus* ("the hanged") in France permits any piece to be moved to *any* vacant cell. This also is believed to end in a draw if played rationally.

Many variations of moving-counter ticktacktoe have been applied to 4 × 4 boards, each player using four counters and striving to get four in a row. A few years ago magician John Scarne marketed an interesting 5 × 5 version called teeko. Players take turns placing four counters each, then alternate with one-unit moves in any direction. A player wins by getting four in a row, orthogonally or diagonally, or in a square formation on four adjacent cells.

Many delightful versions of ticktacktoe do not, however, use moving counters. For example, toetacktick (a name supplied by reader Mike Shodell of Great Neck, New York) is played like the usual game except that the first player to get three in a row *loses*. The second player has a decided advantage. The first player can force a draw only if he plays first in the center. Thereafter, by playing symmetrically opposite the second player, he can ensure the draw.

In recent years, several three-dimensional ticktacktoe games have been marketed. They are played on cubical boards, a win being along any orthogonal or diagonal row as well as on the four main diagonals of the cube. On a 3 × 3 × 3 cube the first player has an easy win. Curiously, the game can never end in a draw because the first player has 14 plays, and it is impossible to make all 14 of them

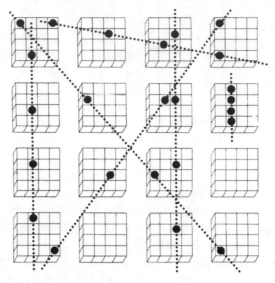

Figure 20. Four-dimensional ticktacktoe. Dotted lines show some winning plays.

without scoring. The $4 \times 4 \times 4$ cube leads to more interesting play and may or may not be a draw if played rationally.

Other ways of playing on cubes have been proposed. Alan Barnert of New York suggests defining a win as a square array of counters on any of the orthogonal planes as well as on the six main diagonal planes. Price Parks and Robert Satten, while students at the University of Chicago in 1941, devised an interesting $3 \times 3 \times 3$ cubical game in which one wins by forming two intersecting rows. The winning move must be on the point of intersection. Because an early move into the center cubicle ensures a win, this move is barred unless it is a winning move or necessary to block an opponent from winning on his next move.

Four-dimensional ticktacktoe can be played on an imaginary hypercube by sectioning it into two-dimensional squares. A $4 \times 4 \times 4 \times 4$ hypercube, for example, would be diagramed as shown in Figure 20. On this board, a win of four in a row is achieved if four marks are in a straight line on any cube that can be formed by assembling four squares in serial order along any orthogonal or either of the two main diagonals. Figure 21 shows a win on such an assembled cube.

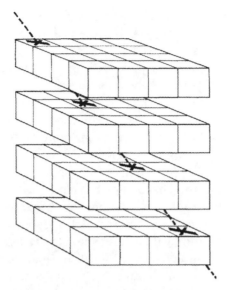

Figure 21. The assembled cube.

The first player is believed to have a sure win, but the game may be a draw if played on a $5 \times 5 \times 5 \times 5$ hypercube. The number of possible rows on which one can win on a cube of n-dimensions is given by the following formula (n is the number of dimensions, k the number of cells on a side):

$$\frac{(k+2)^n - k^n}{2}$$

For an explanation of how this formula is derived, see Leo Moser's comments in the *American Mathematical Monthly*, February 1948, page 99.

The ancient Japanese game of *go-moku* ("five stones"), still popular in Asia, is played on the intersections of a *go* board (this is equivalent to playing on the cells of a 19×19 square). Players take turns placing counters from an unlimited supply until one player wins by getting five in a line, orthogonally or diagonally. No moves are allowed. Experts are of the opinion that the first player can force a win, but as far as I know, no proof of this has ever been published. The game became popular in England in the 1880s under the name of go-bang. It was sometimes played on an ordinary checkerboard, each player using 12 or 15 checkers. Moves were permitted in any direction if no one had won by the time all the checkers were placed.

During the past decade, a number of electrical ticktacktoe-playing machines have been constructed. It is interesting to learn that the first ticktacktoe robot was invented (though never actually built) by Charles Babbage, the nineteenth-century English pioneer inventor of calculating devices. Babbage planned to exhibit his machine in London to raise funds for more ambitious work, but he abandoned his plans after learning that current London exhibits of curious machines (including a "talking machine" and one that made Latin verses) had been financial flops.

A novel feature of Babbage's robot was its method of randomizing choices when faced with alternate lines of equally good play. The machine kept a running total of the number of games won. If called upon to choose between moves A and B, the machine consulted this total, played A if the number was even and B if odd. For three alternatives, the robot divided the total by 3 to obtain a remainder of 0, 1, or 2, each result gearing it to a different move. "It is obvious that any number of conditions might be thus provided for," Babbage writes in his *Passages from the Life of a Philosopher*, 1864, pages 467–471. "An inquiring spectator . . . might watch a long time before he discovered the principle upon which it [the robot] acted."

Unfortunately, Babbage left no record of what he calls the "simple" mechanical details of his machine, so one can only guess at its design. He does record, however, that he "imagined that the machine might consist of the figures of two children playing against each other, accompanied by a lamb and a cock. That the child who won the game might clap his hands whilst the cock was crowing, after which, that the child who was beaten might cry and wring his hands whilst the lamb began bleating." A less imaginative ticktacktoe machine, displayed in 1958 at the Portuguese Industrial Fair in Lisbon, cackled when it won, snarled when it lost (presumably when it was set on a "poor play" circuit).

It might be thought that programming a digital computer to play ticktacktoe, or designing special circuits for a ticktacktoe machine, would be simple. This is true unless your aim is to construct a master robot that will win the maximum number of games against inexperienced players. The difficulty lies in guessing how a novice is most likely to play. He certainly will not move entirely at random, but just how shrewd will he be?

To give an idea of the sort of complications that arise, assume that the novice opens on cell 8. The machine might do well to make an irrational response by seizing cell 3! This would be fatal against an expert, but if the player is only moderately skillful, he is not likely to hit on his one winning reply, cell 9. (See comments on Alain White's article in the bibliography.) Of the six remaining replies, four are disastrous. There will be, in fact, a strong temptation for him to play on cell 4 because this leads to two promising traps against the robot. Unfortunately, the robot can spring its own trap by following with cell 9 and then cell 5 on its next move. It might turn out that in actual play the machine would win more often by this reckless strategy than with a safe course that would most likely end in a draw.

A truly master player, robot or human, would not only know the most probable responses of novices, as determined by statistical studies of past games, but he would also analyze each opponent's style of play to determine what sort of mistakes the opponent would most likely make. If the novice improved as he played, this too would have to be considered. At this point, the humble game of ticktacktoe plunges us into far from trivial questions of probability and psychology.

ADDENDUM

The name "ticktacktoe" has many variations in spelling and pronunciation. According to the *Oxford Dictionary of Mother Goose Rhymes*, 1951, page 406, it derives from an old English nursery rhyme that goes:

> Tit, tat, toe,
> My first go,
> Three jolly butcher boys all in a row.
> Stick one up, stick one down,
> Stick one in the old man's crown.

I have observed that many ticktacktoe players are under the mistaken impression that because they can play an unbeatable strategy they have nothing more to learn about the game. A master player, however, must be quick to take the best possible advantage of a bad play. The following three examples, all from the side opening, will make this clear.

If you open with X8 and your opponent follows with O2, your best response against a novice is X4 because it wins in four out of six moves now open to O. He can block your traps only by playing O7 or O9.

If he opens with X8 and you respond with a lower corner, say O9, you can spring winning traps if he plays X2, X4, or X7.

If he opens with X8, a response of O5 may lead to an amusing development. Should he take X2, you can then permit him to designate your own next move for it is impossible for you to play without being able to set a winning trap!

It was mentioned in the chapter that the moving-counter variation popular in ancient Rome is a win for the first player if he takes the center square. For readers who are interested, the two possible lines of forced play are as follows:

	X	O
	5	3
	4	6
(1)	9	1
	4 *to* 7	*Any move*
	5 *to* 8	
	5	6
	1	9
(2)	3	2
	1 *to* 4	*Any move*
	4 *to* 7	

These lines of play will win regardless of whether moves along the two main diagonals are or are not permitted, but the first one fails if moves along short diagonals are legal.

AFTERWORD, 1988

In the chapter on ticktacktoe, I said that the three-dimensional $4 \times 4 \times 4$ game was unsolved. Oren Patashnik, at Bell Laboratories, cracked the game in 1977 with a computer program almost as complex as the program that proved the four-color map conjecture in 1976. Details are given in Patashnik's article, cited in the bibliography. Incidentally, the proof that the first player can always win a game of Hex if he plays correctly (Chapter 8) also applies to ticktacktoe games. If the game can end in a draw, it proves that the first player can always force either a draw or a win. For more on ticktacktoe and its endless variants, see Chapter 13 in Book 14.

POSTSCRIPT

The magician John Scarne invented and promoted a ticktacktoe-like game in which counters move on a board after they are placed. He called the game teeko, and even published (Crown) an entire book on the game. He was so proud of it that he named a son Teeko Scarne.

By far the most interesting generalization of ticktacktoe was made and analyzed in depth by the noted graph theorist Frank Harary. On the order-3 board of traditional ticktacktoe the objective is to form (or force an opponent to form) a straight tromino. Harary proposed games on $n \times n$ fields in which the objective is *any* polyomino, especially the tetrominoes and pentominoes. I discussed this happy generalization in Chapter 13 of Book 14. The field still poses many intriguing unanswered questions.

Here is a pleasant little problem that I found in the Spring 2007 issue of the British periodical *Mathematical Pie*. Place six Xs on the ticktacktoe field so that there are no three in a line. The solution, which I leave to the reader, is unique.

Can you generalize the problem to $n \times n$ fields? Hint: A main diagonal filled with Os is the smallest set of cells that prevent a row, column, or diagonal from containing n Xs.

BIBLIOGRAPHY

"Tit-tat-to." Alain C. White in *British Chess Magazine*, July 1919. This is the earliest analysis of strategy I have come across. White says deprecatingly that the game can be fully mastered in a half-hour's study. He should have spent another half hour on it, for he makes a whopping error. To the side opening, X8, he recommends O3 as the best reply, failing to see that X9 followed by X5 wins easily.

"Hyper-Spacial Tit-Tat-Toe or Tit-Tat-Toe in Four Dimensions." William Funkenbusch and Edwin Eagle in *National Mathematics Magazine* 19 (December 1944): 119–122.

"Design of a Tit-Tat-Toe Machine." R. Haufe in *Electrical Engineering* 68 (October 1949): 885.

"The Game of Tick-Tack-Toe." Harry D. Ruderman in *Mathematics Teacher* 44 (1951): 344–346.

"Tick-Tack-Toe Computer." Edward McCormick in *Electronics* 25 (August 1952): 154–162.

"Games of Alignment and Configuration." H. J. R. Murray in *A History of Board Games Other Than Chess*, Oxford University Press, 1952, Chapter 3.

Scarne on Teeko. John Scarne. Crown, 1955.

"Relay Moe Plays Tick Tack Toe." Edmund C. Berkeley in *Radio-Electronics* (December 1956).

"Tic-Tac-Toe Mate." David D. Lockhart in *Popular Electronics* (November 1958).

"On Regularity and Positional Games." A. W. Hales and R. I. Jewett in *Transactions of the American Mathematical Society* 106 (1963): 222–229.

The Theory of Gambling and Statistical Logic. Richard A. Epstein. Academic Press, 1967, 359–363.

"The Game of Noughts and Crosses." Fred Schuh in *The Master Book of Mathematical Recreations*, Dover, 1968, Chapter 3.

Your Move. David L. Silverman. McGraw-Hill, 1971, 69–78.

"The Solution of a Simple Game." Daniel I. A. Cohen in *Mathematics Magazine* 45 (September–October 1972): 213–216.

The New Elements of Mathematics. Charles Peirce, ed. Carolyn Eisele. Vol. 2. Humanities Press, 1976, 11–24.

"Tic-Tac-Toe in *n*-Dimensions." Jerome L. Paul in *Mathematics Magazine* 51 (January 1978): 45–49.

"Qubic: 4 × 4 × 4 Tic-Tac-Toe." Oren Patashnik in *Mathematics Magazine* 53 (September 1980): 202–216.

"A Characterization on *n*-Critical Economical Generalized Tic-Tac-Toe Games." X. Liu in *Discrete Mathematics* 110 (1992): 197–203.

"Tic-Tac-Toe on a Finite Plane." Maureen T. Carroll and Steven T. Dougherty in *Mathematics Magazine* 77 (October 2004): 260–274.

Further references are at the end of Chapter 9 in *Wheels, Life, and Other Mathematical Amusements*.

On variants of ticktacktoe, see my Book 5, Chapter 12; Book 7, Chapter 10; Book 6, Chapter 16; Book 10, Chapter 9; and Book 14, Chapter 13.

CHAPTER FIVE

Probability Paradoxes

PROBABILITY THEORY IS A FIELD OF MATHEMATICS unusually rich in paradoxes – truths that cut so strongly against the grain of common sense that they are difficult to believe even after one is confronted with their proofs. The paradox of birth dates is a sterling example. If 24 people are selected at random, what would you estimate the probability to be that two or more of them will have the same birthday (i.e., that is, the same month and day of the year)? Intuitively, you may feel it should be very low. In fact, it is 27/50 or slightly better than 50 percent!

George Gamow, in *One Two Three – Infinity*, gives the following simple method of arriving at this unexpected result. The probability that the birthdays of any two people are *not* alike is clearly 364/365 (since there is only one chance in 365 that one person's birthday will coincide with another's). The probability that a third person's birthday will differ from the other two is 363/365; a fourth person's 362/365, and so on until we reach the twenty-fourth person (342/365). We thus obtain a series of 23 fractions that must be multiplied together to reach the probability that all 24 birthdays are different. The final product is a fraction we approximate by 23/50. In other words, if you were to bet on at least one coincidence of birth dates among 24 people, you would in the long run on average lose 23 and win 27 out of every 50 such bets. (This computation ignores February 29 and also the fact that birth dates tend to be concentrated more in certain months than others; the former lowers the probability, the latter raises it.)

These odds are so surprising that an actual testing of them in a classroom or at a social gathering makes for an entertaining

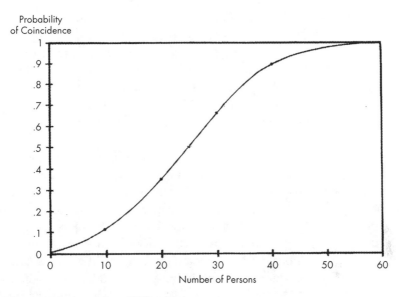

Probability
of Coincidence

Number of Persons

Figure 22. Redrawn from William Ransom.

diversion. If more than 23 people are present, let each person write his birthday on a slip of paper. Collect and compare the slips. More likely than not, at least two dates will match, often much to the astonishment of the parties concerned who may have known each other for years. Fortunately, it does not matter in the least if anyone cheats by giving an incorrect date. The odds remain exactly the same.

An even easier way to test the paradox is by checking birth dates on 24 names picked at random from a *Who's Who* or some other biographical dictionary. Of course the more names you check beyond 24, the greater the probability of a coincidence. Figure 22 (from William R. Ransom's *One Hundred Mathematical Curiosities*, 1955) shows in graph form how the probability curve rises with an increasing number of persons. The graph stops with 60 people because beyond that number the probability is too close to certainty for the curve to be distinguished on the graph from a straight line. Note how the curve climbs steeply until it reaches about 40 people and then levels off toward certainty. For 100 people, the odds for a fair bet on a coincidence are about 3,300,000 to 1. Absolute certainty is not reached, of course, until 366 people are involved.

A neat illustration of the paradox is provided by the birth and death dates of the presidents of the United States. The probability of a coincidence in each case (33 birth dates, 30 death dates) is close to 75 percent. Sure enough, Polk and Harding were born on November 2, and *three* presidents – Jefferson, Adams and Monroe – all died on July 4.

Perhaps even more astounding is the paradox of the second ace. Assume that you are playing bridge and just after the cards are dealt you look over your hand and announce, "I have an ace." The probability that you have a second ace can be calculated precisely. It proves to be 5,359/14,498 which is less than 1/2. Suppose, however, that all of you agree upon a particular ace, say the Ace of Spades. The play continues until you get a hand that enables you to say, "I have the Ace of Spades." The probability that you have another ace is now 11,686/20,825 or slightly *better* than 1/2! Why should naming the ace affect the odds?

The actual computation of chances in these two cases is long and tedious, but the working of the paradox can be easily understood by reducing the deck to only four cards – Ace of Spades, Ace of Hearts, Two of Clubs, and Jack of Diamonds. If these cards are shuffled and dealt to two players, there are only six possible combinations (shown in Figure 23) that a player can hold. Five of these two-card hands permit the player to say, "I have an ace," but in only one instance does he have a second ace. Consequently the probability of the second ace is 1/5. On the other hand, there are only three combinations that permit the player to declare than he holds the Ace of Spades. One of them includes another ace, making the probability of the second ace 1/3.

A similar paradox is that of the second child. Mr. Smith says, "I have two children and at least one of them is a boy." What is the probability that the other child is a boy? One is tempted to say 1/2 until he lists the three possible combinations of equally probable possibilities – BB, BG, GB. Only one is BB, hence the probability is 1/3. Had Smith said that his *oldest* (or tallest, heaviest, etc.) child is a boy, then the situation would be entirely different. Now the combinations are restricted to BB and BG, and the probability that the

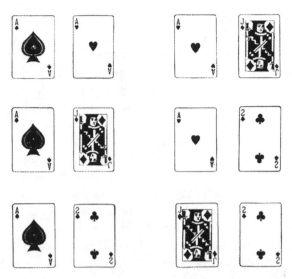

Figure 23

other child is male jumps to 1/2. If this were not the case we would have a most ingenious way to guess the face of a concealed coin with better than even odds. We would simply flip our own coin. If it came up heads, we would reason: "There are two coins here and one of them (mine) is heads. The probability the other is heads is therefore 1/3, so I will bet that it is tails." The fallacy of course is that we are specifying *which* coin is heads. This is the same as identifying the oldest child as the boy, and it changes the odds in a similar fashion.

The most famous of all probability paradoxes is the St. Petersburg paradox, first set forth in a paper by the famous mathematician Daniel Bernoulli before the St. Petersburg Academy. Suppose I toss a penny and agree to pay you a dollar if it falls heads. If it comes up tails, I toss again, this time paying you two dollars if the coin is heads. If it is tails again, I toss a third time and pay four dollars if it falls heads. In short, I offer to double the penalty with each toss and I continue until I am obliged to pay off. What should you pay for the privilege of playing this one-sided game with me?

The unbelievable answer is that you could pay me any amount, say a million dollars, for each game and still expect to come out ahead. In any single game there is a probability of 1/2 that you will win a dollar, 1/4 that you will win two dollars, 1/8 that you will win four dollars, and so on. Therefore the total you may expect to win is $(1 \times 1/2) + (2 \times 1/4) + (4 \times 1/8) + \cdots$. The sum of this endless series is infinite. As a result, no matter what finite sum you paid me in advance per game, you would win in the end if we played enough games. This assumes that I have unlimited capital and that we can play an unlimited number of games. If you paid, say, $1,000 for one game, the odds are high that you would come out a loser. But this expectation is more than balanced by the fact that you have a chance, albeit small, of winning an astronomical sum by a long, unbroken series of tails. If I have only a finite amount of capital, which would always be the case in actual practice, then the fair price for a game is also finite. The St. Petersburg paradox is involved in every "doubling" system of gambling, and its full analysis leads into all sorts of intricate byways.

Carl G. Hempel, a leading figure in the "logical positivist" school and a professor of philosophy at Princeton University, discovered another astonishing probability paradox. Ever since he first explained it in 1937 in the Swedish periodical *Theoria*, "Hempel's paradox" has been a subject of much learned argument among philosophers of science, for it reaches to the very heart of scientific method.

Let us assume, Hempel began, that a scientist wishes to investigate the hypothesis "All crows are black." His research consists of examining as many crows as possible. The more black crows he finds, the more probable the hypothesis becomes. Each black crow can therefore be regarded as a "confirming instance" of the hypothesis. Most scientists feel that they have a perfectly clear notion of what a "confirming instance" is. Hempel's paradox quickly dispels this illusion, for we can easily prove, with ironclad logic, that a purple cow also is a confirming instance of the hypothesis that all crows are black! This is how it is done.

The statement "All crows are black" can be transformed, by a process logicians call "immediate inference," to the logically equivalent statement, "All not-black objects are not-crows." The second statement is identical in meaning with the original; it is simply a different verbal formulation. Obviously, the discovery of any object that confirms the second statement must also confirm the first one.

Suppose then that the scientist searches about for not-black objects in order to confirm the hypothesis that all such objects are not-crows. He comes upon a purple object. Closer inspection reveals that it is not a crow but a cow. The purple cow is clearly a confirming instance of "All not-black objects are not-crows." It therefore must add to the probable truth of the logically equivalent hypothesis, "All crows are black." Of course the same argument applies to a white elephant or a red herring or the scientist's green necktie. As one philosopher recently expressed it, on rainy days an ornithologist investigating the color of crows could continue his research without getting his feet wet. He has only to glance around his room and note instances of not-black objects that are not-crows!

As in previous examples of paradoxes, the difficulty seems to lie not in faulty reasoning but in what Hempel calls a "misguided intuition." It all begins to make more sense when we consider a simpler example. A company employs a large number of typists, some of whom we know to have red hair. We wish to test the hypothesis that all these red-headed girls are married. An obvious way to do this is to go to each red-haired typist and ask her if she has a husband. But there is another way, and one that might even be more efficient. We obtain from the personnel department a list of all unmarried typists. We then visit the girls on this list to check the color of their hair. If none have red hair then we have completely confirmed our hypothesis. No one would dispute the fact that each not-married typist who had not-red hair would be a confirming instance of the theory that the firm's red-headed typists are all married.

We have little difficulty in accepting this investigative procedure because the sets with which we are dealing have a small number of members. But if we are trying to determine whether all crows are black, we have an enormous disproportion between the number of crows on the earth and the number of not-black things. Everyone agrees that checking on not-black things is a highly inefficient way to go about the research. The question at issue is a subtler one – whether it is meaningful to say that a purple cow is in some sense a confirming instance. Does it add, at least in dealing with finite sets (infinite sets lead us into murkier waters), an inconceivably small amount to the probability of our original hypothesis? Some logicians think so. Others are not so sure. They point out, for example, that a purple cow can also be shown, by exactly the same reasoning, to be a confirming instance of "All crows are white." How can an object's discovery add to the probable truth of two contradictory hypotheses?

One may be tempted to dismiss Hempel's paradox with a smile and shrug. It should be remembered, however, that many logical paradoxes that were long regarded as trivial curiosities proved to be enormously important in the development of modern logic. In similar fashion, analyses of Hempel's paradox have already provided valuable insights into the obscure nature of inductive logic, the tool by which all scientific knowledge is obtained.

AFTERWORD

The birth date paradox has been generalized in many ways. I have tried to include the more important papers in my bibliography. William Moser's 1984 article reports the most surprising results. It is hard to believe, but among as few as 14 people, it is more likely than not that two consecutive days will include at least two identical birth dates, and with as few as seven persons, seven consecutive days will also include, with a probability that exceeds one-half, at least two identical birth dates. In researching a paper on puzzles in James Joyce's *Ulysses,* I encountered a surprising birthday coincidence involving Joyce and his good friend, the Irish writer James Stephens. Joyce selected Stephens to complete *Finnegans Wake* in case he (Joyce) died before finishing the book. Both men were born on February 2, 1882.

Hempel's paradox of confirmation theory has become the controversial topic of scores of papers by philosophers and statisticians, not to mention vigorous debates in dozens of books on scientific method. I have tried to list the most important references in the bibliography.

In his *Studies in Deductive Logic* (1884), the British logician and economist William Stanley Jevons poses a problem about an Irishman accused of stealing. To counter the evidence of three witnesses who say they saw him commit the crime, he presents thirty witnesses who swear they did *not* see him stealing. "Where exactly lies the error?" Jevons asks. This always struck me as similar to Hempel's paradox. Is the testimony of the thirty *totally* irrelevant? Or does it add an infinitesimal amount to the plausibility of the Irishman's claim of innocence? Under some circumstances – for example, if one of the thirty saw the Irishman many miles from the scene of the crime at the time of the theft – it would surely add to his credibility.

The literature on Hempel's paradox, which goes to the heart of defining precisely what is a confirming instance of a conjecture, is now vast. I have tried to include the highlights in my bibliography. Here I content myself with quoting my parody of Gelett Burgess's famous quatrain about a purple cow:

> I never saw a purple cow,
> But if I ever see one,
> Will the probability crows are black
> Have a better chance to be 1?

The paradox of the second ace must be stated with extreme care. Both Ball and Littlewood, among others, word the problem so vaguely that it can't be answered. Two provisos must be met to avoid ambiguity. The player who makes the statements must be designated in advance, and the suit of the ace must also be specified. If just any player who holds an ace is allowed to say "I have an ace," the probability of his holding another ace is unaffected by his naming the ace in his hand. Even if a player is specified in advance, if he is allowed to say "I have an ace" without naming the suit, again the probability of his having a second ace is unchanged. In outlining the problem, I met both provisos by calling the player "you" and specifying that the ace was the Ace of Spades. According to Littlewood, this confusing problem goes back to "about 1911."

Note that the probabilities I cited refer to the probabilities of there being *at least* one other ace. The probabilities are not the same if one is concerned with *exactly* one other ace.

Norman Gridgeman, a Canadian statistician, called my attention to the fact that no less eminent a physicist than Erwin Schrödinger, a major architect of quantum mechanics, presented the problem in a 1947 paper. Schrödinger said he first heard of the problem in 1938 from J. H. C. Whitehead, a British mathematician and nephew of Alfred North Whitehead, but according to Littlewood the problem goes back to "about 1911."

Schrödinger met the two provisos as follows: A hand of whist is dealt, but only one player is allowed to pick up his hand. He is asked if he has at least one ace and he truthfully replies, "Yes." He is then asked, "Have you the Ace of Spades?" Again he answers, "Yes." What is the probability that he has at least one other ace? After the first "Yes," the probability is .369+. After the second yes it rises to .561+. How can naming the ace raise the probability from a little more than one-third to more than one-half? Here is Schrödinger's answer:

> It would, of course, have no significance if we had asked A: what is the suit of the ace or one of the aces you are holding, and he had answered: spades. But the fact that among his aces was the one we chose to ask increases the likelihood of his holding more than one. Indeed, the more aces he has, the greater the likelihood of his answering yes to our second question. If a bet were intended, one might call it a rather cunning question.

Gridgeman pointed out in a letter that if the player answers the second question by saying "No, the Ace of Spades is *not* in my hand," this *lowers* the probability of two or more aces to .262+, or slightly better than 1/4.

Observe that all the probabilities cited here refer to the probability of *at least* one other ace. As R. A. Epstein pointed out in a letter (*Scientific American*, July 1957), if one is concerned with *exactly* one more ace, the values are $\frac{8,892}{20,825} = .426+$ for the specified ace, and $\frac{2,223}{7,249} = .306+$ for the unspecified ace. "Also of interest," wrote Epstein, "are the pertinent probabilities if the color of the ace is known. For this condition the probability of one or more aces is .502+, while the probability of one more ace is .403+. Rather oddly, these figures are closer to the specified than the unspecified case."

In a later letter, Gridgeman suggested that the general problem could be usefully displayed by a three-dimensional Venn diagram:

> You could have, say, a cube of unit volume to represent the total sample space (number of possible deals) with, inside, four equal spheres disposed at the corners of a regular tetrahedron, each to be of such a size as to represent the probability of a hand with one ace in it. We label them C, D, H, & S. (Actually, the diagram couldn't be drawn to scale). Now the tetrahedron should be of such a size that the spheres

overlap at the center of each edge (so that the overlaps would represent the probabilities of each of the six possible double-ace hands). At the center of each face the triple overlap of the spheres will represent the probabilities of each of the four possible triple-ace hands. And at the dead center of the tetrahedron all four spheres overlap to represent the probability of a hand with all four aces in it. Get the idea? The probabilities of all kinds of combinations could then be shown. For example, the answer to the question: "If a hand is known to contain two aces, what is the probability of its containing a third?" Or: "If a hand is known to contain two aces, one of which is the ace of spades, what is the probability of its also containing the ace of hearts?"

The problem of the second child also has to be stated with great precision to avoid ambiguity. In Chapter 14 of Book 2, I repeat the problem; I then state it more precisely in Chapter 19.

BIBLIOGRAPHY

THE BIRTHDAY PARADOX

"Understanding the Birthday Problem." Frederick Mosteller in *Mathematics Teacher* 55 (May 1962): 322–325.

"Generalized Birthday Problem." E. H. McKinney in *American Mathematical Monthly* 73 (April 1966): 385–387.

"Extensions of the Birthday Surprise." M. S. Klamkin and D. J. Newman in *Journal of Combinatorial Theory* 3 (October 1967): 279–282.

An Introduction to Probability Theory and Its Applications. William Feller. Vol. 1, 3d ed. Wiley, 1968.

"A Birthday Holiday Problem." Robert Greenwood and Arthur Richert, Jr., in *Journal of Combinatorial Theory* 5 (1968): 422–424.

"A Classroom Illustration of a Nonintuitive Problem." Richard Kleber in *Mathematics Teacher* 62 (May 1969): 361–367.

"More Birthday Surprises." Morton Abramson and W. W. J. Moser in *American Mathematical Monthly* 77 (October 1970): 856–858.

"A Birthday Problem." D. M. Bloom in *American Mathematical Monthly* 80 (December 1973): 1141–1142.

"Another Generalization of the Birthday Problem." J. E. Nymann in *Mathematics Magazine* 48 (January–February 1975): 46–47.

"The Birthday Problem Revisited." Joe Austin in *Two-Year College Mathematics Journal* 7 (January 1976): 39–42.

"A Direct Attack on a Birthday Problem." Samuel Goldberg in *Mathematics Magazine* 49 (May 1976): 130–131.

"Celebrating the Birthday Problem." Neville Spencer in *Mathematics Teacher* 70 (April 1977): 348–353.

"The Uniformity Assumption in the Birthday Problem." Geoffrey Berresford in *Mathematics Magazine* 53 (November 1980): 286–288.

"Using a Microcomputer to Simulate the Birthday Coincidence Problem." John Ginther and William Ewbank in *Mathematics Teacher* 75 (December 1982): 369–370.

"It's Not a Coincidence, but It *Is* a Surprise." William Moser in *Crux Mathematicorum* 10 (1984): 210–213.

"The Birthday Distribution: A Quick Approximation." Anthony Robin in *Mathematical Gazette* 68 (October 1984): 203–206.

"An Extension of the Birthday Problem to Exactly *k* Matches." Robert Hocking and Neil Schwertman in *College Mathematics Journal* 17 (September 1986): 315–321.

"The Birthday Problem for Boys and Girls." Tony Crilly and Shekhar Nandy in *Mathematical Gazette* 71 (March 1987): 19–22.

"On the Generalized Birthday Problem." R. J. McGregor and G. P. Shannon in *Mathematical Gazette* 88 (July 2004): 242–248.

PARADOX OF THE SECOND ACE

A Mathematician's Miscellany. J. E. Littlewood. Methuen, 1953, 27. Cambridge University Press reprint, 1986.

"Aces." George Gamow and Marvin Stern in *Puzzle-Math.* Viking, 1958, 37–42.

Lady Luck. Warren Weaver. Doubleday, 1963, 363. Weaver misunderstood the problem in the first edition, but corrected his remarks in a later printing.

Aha! Gotcha. Martin Gardner. W. H. Freeman, 1982, 105.

"Paradox of the Second Ace." W. W. Rouse Ball and H. S. M. Coxeter in *Mathematical Recreations and Essays.* 13th ed. Dover, 1987, 44–45. The book was originally published in 1882; I do not know in which subsequent edition the paradox first appears.

PARADOX OF THE SECOND CHILD

The Second Scientific American Book of Mathematical Puzzles and Diversions. Martin Gardner. Simon and Schuster, 1961. Rev. ed. University of Chicago Press, 1987, 152, 226.

Aha! Gotcha. Martin Gardner. W. H. Freeman, 1982, 104–105.

HEMPEL'S PARADOX

"Le problème de la vérité." Carl Hempel in *Theoria* 3: 206–246. Göteborg, 1937.

"On Confirmation." Janina Hosiasson-Lindenbaum in *Journal of Symbolic Logic* 5 (1940): 133–148.

"A Purely Syntactical Definition of Confirmation." Carl Hempel in *Journal of Symbolic Logic* 8 (1943): 122–143.

"Studies in the Logic of Confirmation." Carl Hempel in *Mind* 54 (1945): 1–25, 97–121.

"A Note on the Paradoxes of Confirmation." Carl Hempel in *Mind* 55 (1946): 79–83.

Logical Foundations of Probability. Rudolf Carnap. University of Chicago Press, 1950, 223ff, 469ff.

"Hypotheticals." David Pears in *Analysis* 10 (1950): 49–63.

Fact, Fiction, and Forecast. Nelson Goodman. Chapters 3 and 4. Harvard University Press, 1955.

The Logical Problem of Induction. G. H. von Wright. Chapter 6. 2d ed. Macmillan, 1957.

"The Paradoxes of Confirmation." H. G. Alexander in *British Journal for the Philosophy of Science* 9 (1958): 227–233.

"Corroboration versus Induction." J. Agassi in *British Journal for the Philosophy of Science* 9 (1959): 311–317.

"The Paradoxes of Confirmation – A Reply to Dr. Agassi." H. H. Alexander in *British Journal for the Philosophy of Science* 10 (1959): 229–234.

"Popperian Confirmation and the Paradox of the Ravens." David Stove in *Australasian Journal of Philosophy* 37 (1959): 149–151.

"Mr. Stove's Blunders." J. W. N. Watkins in *Australasian Journal of Philosophy* 37 (1959): 240–241.

"On Not Being Gulled by the Ravens." W. J. Huggett in *Australasian Journal of Philosophy* 38 (1960): 48–50.

"A Reply to Mr. Watkins." David Stove in *Australasian Journal of Philosophy* 38 (1960): 51–54.

"Reply to Mr. Stove's Reply." J. W. N. Watkins in *Australasian Journal of Philosophy* 38 (1960): 54–58.

"A Note on Confirmation." Israel Scheffler in *Philosophical Studies* 11 (1960): 21–23.

"Confirmation without Background Knowledge." J. W. N. Watkins in *British Journal for the Philosophy of Science* 10 (1960): 318–320.

"The Paradox of Confirmation." I. J. Good. Parts 1 and 2 in *British Journal for the Philosophy of Science* 11 (1960): 145–149; 12 (1961): 63–64.

"Professor Scheffler's Note." J. W. N. Watkins in *Philosophical Studies* 12 (1961): 16–19.

"Goodman on the Ravens." Sidney Morgenbesser in *Journal of Philosophy* 59 (1962): 493–495.

"The Paradox of Confirmation." J. L. Mackie in *British Journal for the Philosophy of Science* 13 (1963): 265–277. Reprinted in *The Philosophy of Science*, ed. P. H. Nidditch. Oxford University Press, 1968.

The Anatomy of Inquiry. Israel Scheffler. Knopf, 1963, 258–295.

"Confirmation without Paradoxes." William Baumer in *British Journal for the Philosophy of Science* 15 (1964): 177–195.

"Recent Work in Inductive Logic." Henry Kyburg in *American Philosophical Quarterly* 1 (1964): 249–287.

"The Paradoxes of Confirmation." R. H. Vincent in *Mind* 73 (1964): 273–279.

"Confirmation, the Paradoxes, and Positivism." J. W. N. Watkins in *The Critical Approach to Science and Philosophy*, ed. Mario Bunge. Free Press, 1964.

Aspects of Scientific Explanation. Carl Hempel. Free Press, 1965.

"Instantiation and Confirmation." George Schlesinger in *Boston Studies in the Philosophy of Science*, eds. R. S. Cohen and W. W. Wartofsky. Vol. 2. Humanities Press, 1965.

"Hempel and Goodman on the Ravens." David Stove in *Australasian Journal of Philosophy* 43, December (1965).

"Notes on the Paradoxes of Confirmation." Max Black in *Aspects of Inductive Logic*, eds. Jaakko Hintikka and Patrick Suppes. North Holland Publishing Company, 1966. Reprinted in Max Black, *Margins of Precision.* Cornell University Press, 1970.

"A Logic for Evidential Support." L. J. Cohen in *British Journal for the Philosophy of Science* 17 (1966): 105–126.

"Recent Problems of Induction." Carl Hempel in *Mind and Cosmos*, ed. Robert Colodny. University of Pittsburgh Press, 1966.

"Relevance and the Ravens." C. A. Hooker and David Stove in *British Journal for the Philosophy of Science* 18 (1966): 305–315.

"Hempel's Paradox." David Stove in *Dialogue* 4 (1966): 444–455.

"A Bayesian Approach to the Paradoxes of Confirmation." Patrick Suppes in *Aspects of Inductive Logic*, eds. Jaakko Hintikka and Patrick Suppes. North Holland Publishing Company, 1966.

"The Paradoxes of Confirmation." G. H. von Wright in *Aspects of Inductive Logic*, eds. Jaakko Hintikka and Patrick Suppes. North Holland Publishing Company, 1966.

"The White Shoe Is a Red Herring." I. J. Good in *British Journal for the Philosophy of Science* 17 (1967): 322.

"The White Shoe: No Red Herring." Carl Hempel in *British Journal for the Philosophy of Science* 18 (1967): 239–240.

"Confirmation, Qualitative Aspects." Carl Hempel in *The Encyclopedia of Philosophy.* Vol. 2. Macmillan, 1967.

"Baumer on the Confirmation Paradoxes." Howard Kahane in *British Journal for the Philosophy of Science* 18 (1967): 52–56.

"The White Shoe qua Herring Is Pink." I. J. Good in *British Journal for the Philosophy of Science* 19 (1968): 156–157.

"Relevance and The Ravens." C. A. Hooker and D. Stove in *British Journal for the Philosophy of Science* 18 (1968): 305–315.

"On 'Ravens and Relevance' and a Likelihood Solution of the Paradox of Confirmation." L. Gibson in *British Journal for the Philosophy of Science* 20 (1969): 75–80.

"Eliminative Confirmation and Paradoxes." Howard Kahane in *British Journal for the Philosophy of Science* 20 (1969): 160–162.

"The Relevance Criterion of Confirmation." J. L. Mackie in *British Journal for the Philosophy of Science* 20 (1969): 27–40.

"Imagination and Confirmation." Gilbert Walton in *Mind* 78 (1969): 580ff.

The Principles of Scientific Thinking. Rom Harré. Macmillan, 1970, 119–122.

"Theories and the Transitivity of Confirmation." Mary Hesse in *Philosophy of Science* 37 (1970): 50–63.

"Inductive Independence and Confirmation." Jaakko Hintikka in *Essays in Honor of Carl G. Hempel,* ed. Nicholas Rescher. D. Reidel, 1970.

"The Paradoxes of Confirmation – A Survey." R. G. Swinburne in *American Philosophical Quarterly* 8 (1971): 318–330.

Probability and Evidence. A. J. Ayer. Macmillan, 1972.

"Confirmation." Wesley Salmon, in *Scientific American* (May 1973): 75–83.

An Introduction to Confirmation Theory. R. G. Swinburne. Methuen, 1973.

"Logical versus Historical Theories of Confirmation." Alan Musgrave, in *British Journal for the Philosophy of Science* 25 (1974): 1–23.

Confirmation and Confirmability. George Schlesinger. Chapter 1. Oxford University Press, 1974.

"Has Harré Solved Hempel's Paradox?" Nicholas Griffin in *Mind* 84 (1975): 426–430.

"Selective Confirmation and the Ravens." R. H. Vincent in *Dialogue* 14, March (1975).

"Urning a Resolution of Hempel's Paradox." Stuart Meyer in *Philosophy of Science* 44 (1977): 292–296.

"Hempel's Ravens." Martin Gardner in *Aha! Gotcha*. W. H. Freeman, 1982, 133–135.

"The Context of Prediction (and the Paradox of Confirmation)." Tony Lawson in *British Journal for the Philosophy of Science* 36 (1985): 393–407.

"Confirmation, Paradoxes, and Possible Worlds." Shelley Stillwell in *British Journal for the Philosophy of Science* 36 (1985): 19–52.

"Induction and Probability." Martin Gardner in *Time Travel and Other Mathematical Bewilderments*. Chapter 19. W. H. Freeman, 1988.

ST. PETERSBURG PARADOX

"The Application of Probability to Conduct." John Maynard Keynes in *The World of Mathematics*, ed. James Newman. Simon and Schuster, 1956.

Mathematical Recreations and Essays. W. W. Rouse Ball and H. S. M. Coxeter. 13th ed. Dover, 1987, 44–45.

CHAPTER SIX

The Icosian Game and the Tower of Hanoi

TO A MATHEMATICIAN few experiences are more exciting than the discovery that two seemingly unrelated mathematical structures are really closed linked. Recently D. W. Crowe of the University of British Columbia made such a discovery concerning two popular nineteenth-century puzzles: the Icosian Game and the Tower of Hanoi. We shall first describe each puzzle and then show the startling manner in which they are related.

The Icosian Game was invented in the 1850s by the illustrious Irish mathematician Sir William Rowan Hamilton. It was intended to illustrate a curious type of calculus that he had devised and that was similar in many ways to his famous theory of quaternions (the forerunner of modern vector analysis). The calculus could be applied to a number of unusual path-tracing problems on the surfaces of the five Platonic solids, particularly the icosahedron and dodecahedron. Hamilton called it the Icosian calculus, though the game was actually played on the edges of a dodecahedron. In 1859 Hamilton sold the game to a dealer in London for 25 pounds; it was then marketed in several forms in England and on the Continent. This was the only money Hamilton ever received directly, his biographer tells us, for a discovery or publication.

Hamilton suggested a variety of puzzles and games that could be played on the dodecahedron, but the basic puzzle is as follows. Start at any corner on the solid (Hamilton labeled each corner with the name of a large city); then by traveling along the edges make a complete "trip around the world," visiting each vertex once and only once, and return to the starting corner. In other words, the path must form a closed circuit along the edges, passing once through each vertex.

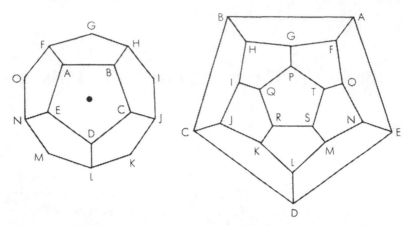

Figure 24. Dodecahedron (left) is punctured (dot) and stretched flat (right). The flat network, which is not in to scale with the solid, is topologically identical with its edges.

If we imagine that the surface of a dodecahedron is made of rubber, we can puncture one of its faces and stretch it open until it lies in a plane. The edges of the surface will now comprise the network shown in Figure 24. This network is topologically identical with the network formed by the edges of the solid dodecahedron, and of course it is much more convenient to handle than the actual solid. The reader may enjoy tackling the round-trip problem on this network, using counters to mark each vertex as it is visited.

On a dodecahedron with unmarked vertices there are only two Hamiltonian circuits that are different in form, one a mirror image of the other. But if the corners are labeled, and we consider each route "different" if it passes through the 20 vertices in a different order, there are 30 separate circuits, not counting reverse runs of these same sequences. Similar Hamiltonian paths can be found on the other four Platonic solids and on many, but not all, semiregular polyhedrons.

The familiar Tower of Hanoi was invented by the French mathematician Edouard Lucas and sold as a toy in 1883. It originally bore the name of "Prof. Claus" of the College of "Li-Sou-Stian," but these were soon discovered to be anagrams for "Prof. Lucas" of the College of "Saint Louis." Figure 25 depicts the toy as it is usually made. The problem is to transfer the tower of eight disks to either

Figure 25. The Tower of Hanoi.

of the two vacant pegs in the fewest possible moves, moving one disk at a time and never placing a disk on top of a smaller one.

It is not hard to prove that there is a solution regardless of how many disks are in the tower and that the minimum number of moves required is expressed by the formula $2^n - 1$ (n being the number of disks). Thus 3 disks can be transferred in 7 moves, 4 in 15, 5 in 31 and so on. For the 8 disks shown in Figure 25, 255 moves are required. The original description of the toy called it a simplified version of a mythical Tower of Brahma in a temple in the Indian city of Benares. This tower, the description read, consists of 64 disks of gold, now in the process of being transferred by the temple priests. When they complete their task, it was said, the temple will crumble into dust and the world will vanish in a clap of thunder. The disappearance of the world may be questioned, but there is little doubt about the crumbling of the temple. The formula $2^{64} - 1$ yields the 20-digit number 18,446,744,073,709,551,615. Assuming that the priests worked night and day, moving one disk every second, it would take them many thousands of millions of years to finish the job.

(The aforementioned number, by the way, is not a prime, but if we increase the number of disks to 89, 107 or 127, the number of moves required to transfer them in each case *is* a prime. They are examples of the so-called Mersenne numbers: primes having

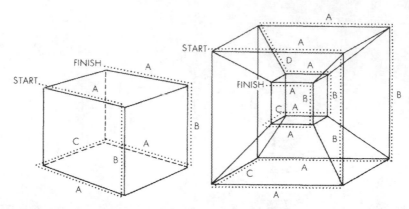

Figure 26. A Hamiltonian path is traced along the edges of a cube at left. The cube has the coordinates A, B, and C; the path follows them in the order ABA-CABA. At right a Hamiltonian path is traced along the edges of a four-dimensional cube projected in three dimensions. This cube has the coordinates A, B, C, and D; the path follows them ABACABADABACABA. This corresponds to the order of transferring four disks in the Tower of Hanoi.

the form of $2^n - 1$. Lucas himself was the first man to verify that $2^{127} - 1$ was a prime. This gargantuan 39-digit number was the largest known prime until 1952, when a large electronic computer was used to find five higher Mersenne primes. Forty-four Mersenne primes are known. The forty-fourth and largest, $2^{32582657} - 1$, was proved prime in 2006. It has 9,808,358 digits.

A Tower of Hanoi puzzle is easily made by cutting eight cardboard squares of graduated sizes (or using playing cards from the ace to the eight) and moving them among three spots on a piece of paper. If the spots form a triangle, the following simple procedure will solve the puzzle for any number of "disks." Transfer the smallest disk on every other play, always moving it around the triangle in the same direction. On the remaining plays, make the only transfer possible that does not involve the smallest disk.

How is this puzzle related to Hamilton's game? To explain the connection we must first consider a tower of three disks only, labeling the disks, from top to bottom, A, B, and C. If we follow the procedure given above, we solve the puzzle by moving the disks in the following order: ABACABA.

Let us now label with A, B, and C the three coordinates of a regular hexahedron, commonly called a cube (see illustration at left of Figure 26). If we trace a path along the edges of the cube, choosing the coordinates in the order ABACABA, the path will form a Hamiltonian circuit! Crowe saw that this could be generalized as follows: The order of transferring n disks in the Tower of Hanoi puzzle corresponds exactly to the order of coordinates in tracing a Hamiltonian path on a cube of n dimensions.

An additional illustration will make this clear. Although we cannot make a model of a four-dimensional cube (called a hypercube or tesseract), we can project the network of its edges in the three-dimensional model depicted at right of Figure 26. This network is topologically identical to the network of edges on a hypercube. We label its coordinates A, B, C, and D, the D coordinate being represented by the diagonal lines.

The order for transferring a tower of four disks is ABACABADABA-CABA. When we traverse the hypercube model, making our turns correspond to this sequence, we find ourselves tracing a Hamiltonian path. By the same token, five disks transfer in an order corresponding to a Hamiltonian circuit on a five-dimensional hypercube, six disks correspond to a six-dimensional hypercube, and so on.

ADDENDUM

Proving that n disks in the Tower of Hanoi can be moved to another peg in $2^n - 1$ steps is not difficult and is an excellent classroom exercise in mathematical induction. (See *Mathematics Teacher*, 1951, Vol. 44, page 505; and 1952, Vol. 45, page 522.) The puzzle is easily generalized to any number of pegs. (See Henry Ernest Dudeney's *The Canterbury Puzzles*, 1907, Problem No. 1; and the *American Mathematical Monthly*, March 1941, Problem No. 3918.)

The isomorphism of the Tower of Hanoi's solution and the Hamiltonian path on cubes and hypercubes is perhaps not so startling when we realize that in both cases the sequence of moves is a pattern familiar to anyone working with binary computers. We first write the binary numbers from 1 to 8 and label the columns A, B, C, D as shown in Figure 27. We then write opposite each row the letter that identifies the "1" that is farthest to the right on each row. The sequence of these letters from top down will be the pattern in question.

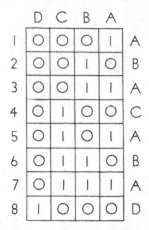

Figure 27. Table of binary numbers.

This pattern is encountered frequently in mathematical puzzles. Cards for guessing a thought-of number and an ancient mechanical puzzle called the Chinese rings are two additional examples. The most familiar instance of the pattern is the sequence in the sizes of marks on a one-inch segment of an ordinary ruler (see Figure 28). The pattern results, of course, from successive binary divisions of the inch into halves, quarters, eighths, and sixteenths.

POSTSCRIPT

Finding Hamiltonian paths on a graph can be the basis of many puzzles. Here (Figure 29) is one from Mark Lynch's article "Creating Recreational Hamiltonian Cycle Problems," in the *Mathematical Gazette*, July 2004, Vol. 88, pp. 215–218.

Figure 28. Binary divisions of an inch.

Figure 29. (Artist: Harold Jacobs)

There is only one Hamiltonian cycle. I won't spoil the pleasure of finding it by giving the solution.

The Tower of Hanoi can be solved by applying what is called a Gray code. See the chapter on binary Gray codes in my Book 11 for the tower's isomorphism with a Chinese rings puzzle, and other mechanical puzzles.

The Tower of Hanoi has been marketed many times around the world and in many different forms. An unusual version, sold in the United States in 1974, consisted of cardboard forms that folded to make nine pyramids of nine different sizes. One was told to hide a small object, such as a ring or coin, under the smallest pyramid and then to stack the other eight pyramids on top. The task was to obtain the object by moving the pyramids back and forth.

Two-person games have been based on the puzzle. Harry Weller-ton, in the British magazine *Games and Puzzles* (December 1976), proposed this one. The board is a row of seven squares. Each player has, at one end of the row, five disks stacked in sequential order, largest disk on the bottom. The stacks are of two different colors. Players alternate moving a disk to another square. A disk may be moved to an empty square or to another disk (of either color) provided it is larger than the disk moved. A disk arriving on the opponent's cell cannot be moved again. The winner is the first to get all his disks, in proper order, on the opponent's starting cell.

Many readers rediscovered an old technique for solving the traditional puzzle. If disks at even positions in the initial stack are one color, and those in odd positions are of another color, the solving procedure is simple. Disks of one color go around the triangle in one direction. Those of the other color go the other way. The procedure is described by R. E. Allardice and A. Y. Fraser in "Le Tour d' Hanoi," *Proceedings of the Edinburgh Mathematical Society*, Vol. 2, 1884, pp. 50–53.

The Tower of Hanoi obviously generalizes to n disks and k pegs. The task is to transfer the disks in the smallest number of moves to another peg. Henry Dudeney considered the $k = 4$ case in *The Weekly Dispatch* (May 25, 1902), and the $k = 5$ case in the same newspaper for March 15, 1903. He deals with the general problem in *The Canterbury Puzzles* where he gives a recursive procedure for calculating what he believed to be the minimum number of moves for any n, provided n is a triangular number when $k = 4$, or a pyramidal number when $k = 5$. For example, consider four pegs and ten disks (10 is a triangular number). The minimum number of moves, 49, is considerably less than the $2^{10} - 1 = 1,023$ moves needed if there are only three pegs. By adding one more peg, the priests at Benares could transfer all 64 disks in just a few hours!

The general case was posed as Problem 3918 in the *American Mathematical Monthly* in 1939, and two partial solutions were given in Vol. 48, 1941, pp. 216–219. Since then many mathematicians have struggled with the general task, including Donald Knuth, Ashok Chandra, Don Oliver, and Glen Manacher. All confirmed Dudeney's formulas. Unfortunately, all make the following assumption. Disks that are adjacent on a peg are said to have a "gap" if they are not consecutive in size. The assumption is that any solution allowing gaps would be longer than one without gaps.

No shorter solution has been found for any values of n and k that violate this assumption, but the general problem remains unsolved in spite of many claims to the contrary. What is required is a proof that no solution violating the assumption is shorter. Even the case of $k = 4$ is unsolved for all but small values of n.

The Tower of Hanoi is discussed in *Concrete Mathematics* (Addison-Wesley, 1994), by Ronald Graham, Donald Knuth, and Oren Patashnik, and Knuth will consider the general problem in the forthcoming next volume in his famous series on *The Art of Computer Programming*.

The Tower of Hanoi has been modified by other constraints. R. S. Scorer, P. M. Grundy, and C. A. B. Smith, in a 1944 paper, proposed putting the pegs in a straight line and not allowing moves from one end peg to the other. With this restriction, $3^n - 1$ moves are needed to transfer the disks from one end to the other. The sequence of moves corresponds to a ternary Gray code. After $\frac{3^n-1}{2}$ moves, the stack is halfway home on the middle peg.

Dozens of other variants have been proposed, such as moving the pegs around a triangle only in one direction. Scorer, Grundy, and Smith showed that $3^n - 1$ moves suffice in this case as well, when $k = 4$ and when the stack is supposed to move two steps. For example, if there are $n = 3$ disks numbered 0, 1, 2 from smallest to largest, their 26-move solution is 00100100 2 00100100 2 00100100. But Knuth noticed in 1975 that this isn't optimum. He found the 18-move solution 000112 012010 101000. (This incidentally must be the best, because there must be at least 4 more 0's than 1's and 4 more 1's than 2's.)

The cyclic problem, like the unrestricted problem, remains unsolved for general n when $k > 3$, although the complete solution is known in the three-peg case. M. D. Atkinson proved in 1981 that for n disks and $k = 3$ the minimum number of clockwise moves required is exactly

$$\frac{(1 + \sqrt{3})^{n+1} - (1 - \sqrt{3})^{n+1}}{2\sqrt{3}} - 1$$

For $n = 1$ through $7 (k = 3)$ the minimum number of clockwise moves for transferring n disks to the next adjacent peg clockwise are 1, 5, 15, 43, 119, 327, and 895. For clockwise transferring to the next adjacent peg counterclockwise, 2, 7, 21, 59, 163, 447, and 1,223 moves are required.

BIBLIOGRAPHY

THE ICOSIAN GAME

"Sir William Hamilton's Icosian Game." A. S. Herschel in *Quarterly Journal of Applied Mathematics* 5 (1862): 305.

The Life of Sir William Rowan Hamilton. Robert Graves. Vol. 3. Dublin, 1882–1889, 55ff.

THE TOWER OF HANOI

The Canterbury Puzzles. H. E. Dudeney. Thomas Nelson, 1919. Dover reprint, 1958.

"Some Binary Games." R. S. Scorer, P. M. Grundy, and C. A. B. Smith in *Mathematical Gazette* 28 (1944): 96–103.

"The *n*-Dimensional Cube and the Tower of Hanoi." D. W. Crowe in *American Mathematical Monthly* 63 (January 1956): 29–30.

The Masterbook of Mathematical Recreations. Frederick Schuh. Dover, 1968, 119–121.

"The Tower of Brahma Revisited." Ted Roth in *Journal of Recreational Mathematics* 7 (Spring 1974): 116–119.

"Tower of Hanoi with More Pegs." Brother Alfred Brousseau in *Journal of Recreational Mathematics* 8 (1975–1976): 169–176. This and the preceding article are reprinted in *Mathematical Solitaires and Games*, ed. Benjamin Schwartz. Baywood, 1980.

Combinatorial Algorithms: Theory and Practice. E. M. Reingold, J. Nievergelf, and N. Deo. Prentice-Hall, 1977, 29–30.

"Another Look at the Tower of Hanoi." Michael Shwarger in *Mathematics Teacher* (September 1977): 528–533.

"The Cyclic Tower of Hanoi." M. D. Atkinson in *Information Processing Letters* 13 (1981): 118–119.

"The Towers of Brahma and Hanoi Revisited." Derick Wood in *Journal of Recreational Mathematics* 14 (1981–1982): 17–24.

"Computer Recreations." A. K. Dewdney in *Scientific American* (November 1984): 19–28.

"Towers of Hanoi and Analysis of Algorithms." Paul Cull and E. F. Ecklund, Jr., in *American Mathematical Monthly* 92 (June–July 1985): 407–420.

"The Binary Gray Code." Martin Gardner in *Knotted Doughnuts and Other Mathematical Entertainments.* Chapter 2. W. H. Freeman, 1986.

Mathematical Recreations and Essays. W. W. Rouse Ball and H. S. M. Coxeter. 13th ed. Dover, 1987, 316–317.

"All You Ever Wanted to Know about the Tower of Hanoi but Were Afraid to Ask." W. F. Lunnon. Unpublished typescript.

CHAPTER SEVEN

Curious Topological Models

Why did the chicken cross the Moebius strip?
Answer: To get to the same side.

AS MANY READERS OF THIS BOOK ARE AWARE, a Moebius band is a geometrical curiosity that has only one surface and one edge. Such figures are the concern of the branch of mathematics called topology. People who have a casual interest in mathematics may get the idea that a topologist is a mathematical playboy who spends his time making Moebius bands and other diverting topological models. If they were to open any recent textbook of topology, they would be surprised. They would find page after page of symbols, seldom relieved by a picture or diagram. It is true that topology grew out of the consideration of geometrical puzzles, but today it is a jungle of abstract theory. Topologists are suspicious of theorems that must be visualized in order to be understood.

Serious topological studies nonetheless produce a constant flow of weird and amusing models. Consider, for example, the double Moebius band. This is formed by placing two strips of paper together, giving them a single half-twist as if they were one strip, and joining their ends as shown in Figure 30.

We now have what appears to be two nested Moebius bands. Indeed, you can "prove" that there are two separate bands by putting your finger between the bands and running it all the way around them until you come back to the point at which you started. A bug crawling between the bands could circle them indefinitely, always walking along one strip with the other strip sliding along its back. At no point would he find the "floor" meeting the "ceiling."

Figure 30. Double Moebius band is made by placing two strips of paper together (left), giving both of them a half-twist and joining their ends as indicated at right. (Artist: James Egleson.)

An intelligent bug would conclude that he was walking between the surfaces of two separate bands.

Suppose, however, that the bug made a mark on the floor, and circled the bands until he reached the mark again. It would find the mark not on the floor but on the ceiling, and it would require a second trip around the bands to find it on the floor again! The bug would need considerable imagination to comprehend that both floor and ceiling were one side of a single strip. What appears to be two nested bands is actually one large band! Once you have opened the model into the large band, you will find it a puzzling task to restore it to its original form.

When the band is in its double form, two separate edges of it run parallel to each other; they circle the model twice. Imagine that these edges are joined and that the band is made of thin rubber. You would then have a tube that could be inflated to make a torus (the topologist's term for the surface of a doughnut). The joined edges would form a closed curve that coiled twice around the torus. This means that a torus can be cut along such a curved line to form the double Moebius band.

The double band is identical, in fact, with a single band that is given four half-twists before its edges are joined. It is possible to cut a torus into a band with any desired even number of half-twists, but impossible to cut it so as to produce bands with an odd number of such twists. This is because the torus is a two-sided surface and only bands with an even number of half-twists are two-sided. Although two-sided surfaces can be made by cutting one-sided ones, the

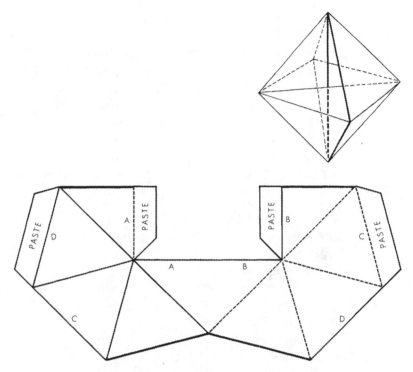

Figure 31. Moebius band with triangular edge was devised by Bryant Tuckerman. If the figure at bottom is redrawn, preferably on a larger scale, the polyhedral model at upper right may be assembled as follows. First, cut out the figure. Second, fold it "down" along the solid lines. Third, fold it in the opposite direction along the broken lines. Fourth, by applying paste to the four tabs, join edges A and A, B and B, C and C, and D and D. The heavy lines in the finished polyhedron trace the triangular boundary of the Moebius surface. (Artist: James Egleson.)

reverse is not possible. If we wish to obtain one-sided bands (bands with an odd number of half-twists) by cutting a surface without edges, we must resort to cutting a Klein bottle. The Klein bottle is a closed one-sided surface with no edges; it can be bisected into two Moebius strips that are mirror images of each other.

The simple Moebius band is made by giving a strip one half-twist before joining the ends. Can the band somehow be stretched until this edge is a triangle? The answer is "yes." The first man to devise such a model was Bryant Tuckerman, one of the four pioneers in the art of folding flexagons (see Chapter 1). Figure 31 shows how a piece of paper can be cut, folded, and pasted to create Tuckerman's model.

Surfaces may not only have one or two sides; they may also differ topologically in the number and structure of their edges. Such traits cannot be altered by distorting the surface; hence they are called topological invariants. Let us consider surfaces with no more than two edges, and edges that are either simple closed curves or in the form of an ordinary single knot. If the surface has two edges, they may be independent of each other or linked. Within these limits, we can list the following 16 kinds of surfaces (excluding edgeless surfaces such as the sphere, the torus, and the Klein bottle):

ONE-SIDED, ONE-EDGED
  1. Edge is a simple closed curve.
  2. Edge is knotted.
TWO-SIDED, ONE-EDGED
  3. Edge is a simple closed curve.
  4. Edge is knotted.
ONE-SIDED, TWO-EDGED
  5. Both edges are simple closed curves, unlinked.
  6. Both edges are simple closed curves, linked.
  7. Both edges are knotted, unlinked.
  8. Both edges are knotted, linked.
  9. One edge is simple; one knotted, unlinked.
  10. One edge is simple; one knotted, linked.
TWO-SIDED, TWO-EDGED
  11. Both edges are simple closed curves, unlinked.
  12. Both edges are simple closed curves, linked.
  13. Both edges are knotted, unlinked.
  14. Both edges are knotted, linked.
  15. One edge is simple; one knotted, unlinked.
  16. One edge is simple; one knotted, linked.

Paper models are easily constructed to illustrate examples of each of these 16 surfaces. Models for surfaces 1 through 12 are depicted in Figure 32. Models of the remaining four surfaces are shown in Figure 33.

When some of these models are cut with scissors in certain ways, the results are startling. As almost everyone who has played with

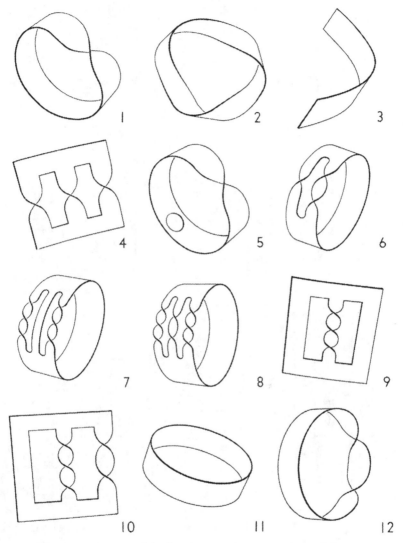

Figure 32. Paper models of surfaces 1 to 12. (Artist: James Egleson.)

a Moebius band knows, cutting the band in half lengthwise does not produce two separate bands, as one might expect, but one large band. (The large band has four half-twists; thus it can be made up into the double Moebius band described earlier.) Not so well-known is the fact that if you start the cut a third of the way between one

Figure 33. Paper models of surfaces 13 to 16. (Artist: James Egleson.)

edge and the other, and cut until you return to the starting point, the Moebius band opens into a large band linked with a smaller one.

Cutting surface 12 in half yields two interlocked bands of the same size, each exactly like the original one. Cutting surface 2 in half results in a large band that has a knot in it. This latter stunt was the

Figure 34. Interlocked curves that can be separated without passing one through an opening in the other. The curves at the top may be separated by passing the twisted curve through itself at A. (Artist: James Egleson.)

subject of a booklet that enjoyed a wide sale in Vienna in the 1880s. The booklet revealed the secret of forming a knot in a cloth band without resorting to magical trickery.

In saying that two edges are "linked," we mean linked in the manner of two links in a chain. To separate the links, it is necessary to open one link and pass the other through the opening. It is possible, however, to interlock two closed curves in such a manner that in order to separate them it is not necessary to pass one through an opening in the other. The simplest way to do this is shown by the upper curves in Figure 34. These curves can be separated by passing one band through *itself* at point A.

The three closed curves at the bottom of the illustration also are inseparable without being linked. If you remove any one curve, the other two are free; if you link any pair of curves, it frees the third one. This structure, by the way, is topologically identical with the familiar three-ring trademark of a well-known brand of beer. These rings are sometimes called Borromean rings because they formed the coat of arms for the Renaissance Italian family of Borromeo. I know of no paper model of a single surface, free from self-intersection, that has two or more edges locked without being linked, but perhaps a clever reader can succeed in constructing one.

ADDENDUM

An interesting model of the double Moebius band can be made of rigid plastic. This makes it easy for someone to run his finger all the way around between the "two" bands.

Mel Stover of Winnipeg wrote that he made a model in flexible white plastic and then inserted a strip of red plastic between "them." Since the red strip is clearly seen to be at all points between what appear to be two separate bands, the surprise is heightened when the red band is slipped out and the white strip shown to be a single band. The red strip must have open ends that are overlapped rather than joined; otherwise, it will be linked to the white band and cannot be slipped out.

The red strip in Stover's model, when it is placed within the white strip, assumes the form of a Moebius band. Every non-orientable (one-sided) surface can be covered in a similar fashion by what has been called a "two-sheeted" bilateral surface. For example, the Klein bottle can be covered completely by a torus, half of which must be turned inside out. Like the Moebius strip covering, this surface appears to be two separate surfaces, one within the other. If you puncture it at any point, you find the inner surface separated from the outer by the surface of the Klein bottle, yet the inner and outer surfaces are parts of the same torus. (See *Geometry and the Imagination* by David Hilbert and S. Cohn-Vossen, English translation, 1956, page 313.)

POSTSCRIPT

Modern topology, which began with the work of Poincaré, has become one of the most active branches of today's mathematics. The number of textbooks is now in the hundreds. More than 15 paperbacks on topology are available as Dover reprints. Topology has become increasingly essential for understanding particle physics, especially superstring theory or M-theory as it is now called.

The Moebius band and the closely related Klein bottle turn up in many later *Scientific American* columns, especially in Chapter 2 of Book 5, and Chapter 9 of Book 7. Solid analogs of Moebius bands are treated in Chapter 5 of Book 14. A collection of my short stories titled *The No-Sided Professor* (Prometheus, 1987) reprints a tale of the same title. It tells how a topologist discovered a way to fold and twist a piece of paper so it becomes *no-sided* and vanishes with a pop. On the Borromean rings see Book 4, Chapter 2.

BIBLIOGRAPHY

"A Non-Singular Polyhedral Möbius Band Whose Boundary Is a Triangle." Bryant Tuckerman in *American Mathematical Monthly* 55 (May 1948): 309–311.

"Topology." Albert W. Tucker and Herbert S. Bailey, Jr., in *Scientific American* 182 (January 1950): 18–24.

Mathematical Models. H. Martyn Cundy and A. P. Rollett. Clarendon Press, 1952.

Intuitive Concepts in Elementary Topology. Bradford Arnold. Prentice-Hall, 1962.

Experiments in Topology. Stephen Barr. Crowell, 1964.

Visual Topology. W. Lietzmann. Chatto and Windus, 1965.

"Klein Bottles and Other Surfaces." Martin Gardner in *The Sixth Book of Mathematical Games from Scientific American.* Chapter 2. University of Chicago Press, 1983.

"No-Sided Professor." Martin Gardner in *The No-Sided Professor and Other Tales of Fantasy, Humor, Mystery, and Philosophy.* Prometheus Books, 1987.

Topology Now! Robert Messer and Philip Straffin. Mathematical Association of America, 2006.

CHAPTER EIGHT

The Game of Hex

IT IS SOMETHING OF AN OCCASION these days when someone invents a mathematical game that is both new and interesting. Such a game is Hex, introduced 15 years ago at Niels Bohr's Institute for Theoretical Physics in Copenhagen. It may well become one of the most widely played and thoughtfully analyzed new mathematical games of the century.

Hex is played on a diamond-shaped board made up of hexagons (see Figure 35). The number of hexagons may vary, but the board usually has 11 on each edge. Two opposite sides of the diamond are labeled "black"; the other two sides are "white." The hexagons at the corners of the diamond belong to either side. One player has a supply of black pieces; the other, a supply of white pieces. The players alternately place one of their pieces on any one of the hexagons, provided the cell is not already occupied by another piece. The objective of Black is to complete an unbroken chain of black pieces between the two sides labeled "black." White tries to complete a similar chain of white pieces between the sides labeled "white."

The chain may freely twist and turn; an example of a winning chain is shown in Figure 35. The players continue placing their pieces until one of them has made a complete chain. The game cannot end in a draw, because one player can block the other only by completing his own chain. These rules are simple, yet Hex is a game of surprising mathematical subtlety.

Hex was invented by Piet Hein, who must surely be one of the most remarkable men in Denmark. Piet Hein began his career as a student at the Institute for Theoretical Physics; then his industrial inventions switched him to engineering, where he remained until the Germans invaded Denmark in 1940. Because Piet Hein was

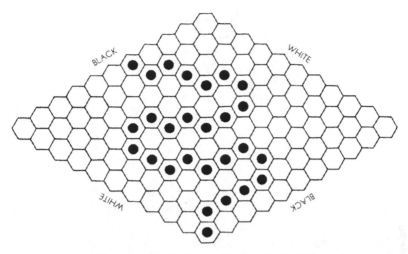

Figure 35. A winning chain for Black on a Hex board with 11 hexagons on each side. (Artist: Amy Kasai.)

the head of an anti-Nazi group, he was forced to go underground. After the war, he became well-known as a writer on scientific and other topics for *Politiken*, the leading Danish newspaper. He is also known, under the pseudonym of Kumbel, as the author of numerous volumes of epigrammatical poems. These books have sold in the millions.

The game of Hex occurred to Piet Hein while he was contemplating the famous four-color theorem of topology. (The theorem, proved in 1976, is that four colors are sufficient to make any map so that no two countries of the same color have a common boundary.) Piet Hein introduced the game in 1942 with a lecture to students at the Institute. On December 26 of that year, *Politiken* published an account of the game; it soon became enormously popular in Denmark under the name of Polygon. Pads on which the game could be played with a pencil were sold, and for many months *Politiken* ran a series of Polygon problems, with prizes for the best solutions.

In 1948 John F. Nash, then a graduate student in mathematics at Princeton University (later a professor at Massachusetts Institute of Technology and one of the nation's outstanding authorities on game theory), independently reinvented the game. It quickly captivated

Figure 36. (Artist: Amy Kasai.)

students of mathematics both at the Institute for Advanced Study and Princeton. The game was commonly called either Nash or John, the latter name referring mainly to the fact that it was often played on the hexagonal tiles of bathroom floors. It did not acquire the name Hex until 1952 when a version of the game was issued under that title by Parker Brothers, Inc.

Readers who would like to try Hex are advised to make photocopies of the board. The game can be played on these sheets by marking the hexagons with circles and crosses. If you should prefer to play with removable pieces on a permanent board, a large one can easily be drawn on heavy cardboard or made by cementing together hexagonal tiles. If the tiles are big enough, ordinary checkers make convenient pieces.

One of the best ways to learn the subtleties of Hex is to play the game on a field with a small number of hexagons. When the game is played on a 2×2 board (four hexagons), the player who makes the first move obviously wins. On a 3×3 board the first player wins easily by making his first move in the center of the board (see Figure 36). Because Black has a double play on both sides of his piece, there is no way in which his opponent can keep him from winning on his third move.

On a 4×4 board things begin to get complicated. The first player is sure to win if he immediately occupies any one of the four cells numbered in Figure 37. If he makes his opening play elsewhere, he can always be defeated. An opening play in cell 2 or 3 ensures a win on the fifth move; an opening play in cell 1 or 4, a win on the sixth move.

On a 5×5 board it can still be shown that if the first player immediately occupies the hexagon in the center, he can win on his

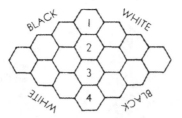

Figure 37. (Artist: Amy Kasai.)

seventh move. On larger fields, the analysis becomes enormously difficult. Of course the standard 11×11 board introduces such an astronomical number of complications that a complete analysis seems beyond the range of human computation.

Game theorists find Hex particularly interesting for the following reason. Although no "decision procedure" is known that will ensure a win on a standard board, there is an elegant *reductio ad absurdum* "existence proof" that there is a winning strategy for the first player on a field of any size! (An existence proof merely proves the existence of something without telling you how to go about finding it.) The following is a highly condensed version of the proof (it can be formulated with much greater rigor) as it was worked out in 1949 by John Nash:

1. Either the first or second player must win; therefore, there must be a winning strategy for either the first or second player.
2. Let us assume that the second player has a winning strategy.
3. The first player can now adopt the following defense. He first makes an arbitrary move. Thereafter he plays the winning second-player strategy assumed initially. In short, he becomes the second player, but with an extra piece placed somewhere on the board. If in playing the strategy he is required to play on the cell where his first arbitrary move was made, he makes another arbitrary move. If later he is required to play where the second arbitrary move was made, he makes a third arbitrary move, and so on. In this way, he plays the winning strategy with one extra piece always on the field.

Figure 38. Three problems of Hex. (Artist: Amy Kasai.)

4. This extra piece cannot interfere with the first player's imitation of the winning strategy, for an extra piece is always an asset and never a handicap. Therefore the first player can win.

5. Since we have now contradicted our assumption that there is a winning strategy for the second player, we are forced to drop this assumption.

6. Consequently, there must be a winning strategy for the first player.

There are a number of variations on the basic theme of Hex, including a version in which each player tries to force his opponent to make a chain. According to a clever proof devised by Robert Winder, a graduate student of mathematics at Princeton, the first player can always win this game on a board that has an even number of cells on a side, and the second player can always win on a board with an odd number.

After the reader has played Hex for a while, he may wish to tackle three problems devised by Piet Hein. These are set forth in the three illustrations of Figure 38. The objective in all three problems is to find the first move that will ensure a win for White.

ADDENDUM

Hex can be played on several different types of fields that are topologically equivalent to the fields of hexagons. A field of equilateral triangles, for example, may be used, placing the counters on the intersections. An ordinary checkerboard is isomorphic with a Hex field if one assumes that the squares connect diagonally in

Figure 39. How second player pairs the cells to win on a "short" board. (Artist: Amy Kasai.)

one direction only (say, NE and SW, but not NW and SE). Both boards seem to me less satisfying for actual play than the mosaic of hexagons.

Several shapes for a Hex field other than the diamond have been proposed. For example, John Milnor, of Princeton University, suggested a field in the shape of an equilateral triangle. The winner is the first to complete a chain connecting all three sides of the triangle. Corner cells are regarded as belonging to both their adjacent sides. Nash's proof of first-player-win applies with equal force to this variant.

To counter the strong advantage held by the first player in the standard game of Hex, several proposals have been made. The first player may be forbidden to open on the short diagonal. The winner may be credited with how few moves it took him to win. The first player opens with one move, but thereafter each player has two moves per turn.

It is tempting to suppose that on an n by $n + 1$ board (e.g., a 10×11), with the first player taking the sides that are farthest apart, the relative advantages of the two players might be made more equal. Unfortunately, Nash discovered a simple strategy that gives the second player a certain win. The strategy involves a reflection symmetry along a central axis. If you are the second player, you imagine the cells to be paired according to the scheme indicated by the letters in Figure 39. Whenever your opponent plays, you play on the other cell with the same letter. Owing to the shorter distance

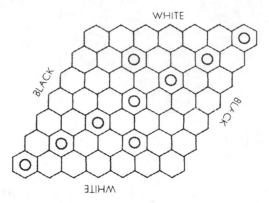

Figure 40. (Artist: Amy Kasai.)

between your two sides of the board, it is impossible for you to lose!

A few words about general strategy in playing Hex. Quite a number of readers wrote that they were disappointed to discover that the first player has an easy win simply by taking the center cell, then extending a chain of adjacent cells toward his two sides of the board. They argued that since he always has a choice of two cells for the next link in the chain, it would be impossible to block him. Of course they failed to play long enough to discover that chains can be blocked by taking cells that are not adjacent to the ends of the chain. The game is much subtler than it first appears. Effective blocking often involves plays that seem to have no relationship to the chain that is being blocked.

A more sophisticated strategy is based on the following procedure. Play first in the center; then seek to form on each of your sides a chain of separated links that are either diagonal or vertical, like the two chains shown in Figure 40. If your opponent checks you vertically, you switch to a diagonal play, and if he checks you diagonally, you switch to vertical. Of course, once you succeed in joining your two sides with a disconnected chain on which each missing link is a double play, you cannot be blocked. This is a good strategy to play on novices, but it can be countered by proper defensive moves.

Still another strategy provided the basis of a Hex machine constructed by Claude Shannon and E. F. Moore, both at that time on the staff of Bell Telephone Laboratories. Here is Shannon's description of the device from his article on "Computers and Automata" in the *Proceedings of the Institute of Radio Engineers*, Vol. 41, October 1953:

> After a study of this game, it was conjectured that a reasonably good move could be made by the following process. A two-dimensional potential field is set up corresponding to the playing board, with white pieces as positive charges and black pieces as negative charges. The top and bottom of the board are negative and the two sides positive. The move to be made corresponds to a certain specified saddle point in this field.
>
> To test this strategy, an analog device was constructed, consisting of a resistance network and gadgetry to locate the saddle points. The general principle, with some improvements suggested by experience, proved to be reasonably sound. With first move, the machine won about seventy per cent of its games against human opponents. It frequently surprised its designers by choosing odd-looking moves that, on analysis, proved sound. We normally think of computers as expert at long, involved calculations and poor in generalized value judgments. Paradoxically, the positional judgment of this machine was good; its chief weakness was in end-game combinatorial play. It is also curious that the Hex-player reversed the usual computing procedure in that it solved a basically digital problem by an analog machine.

As a joke, Shannon also built a Hex machine that took the second move and always won, much to the puzzlement of Hexperts who knew of the first player's strong advantage. The board was short in one direction (7 × 8), but mounted on a rectangular box in such a way that the inequality of sides was disguised. Few players were suspicious enough to count the cells along two edges. The machine, of course, played the winning reflection strategy previously described. It could have been constructed to respond instantly to moves, but thermistors were used to slow down its operation. It took one to eight seconds to reach a decision, thus conveying the impression that it was making a complicated analysis of the configuration on the field!

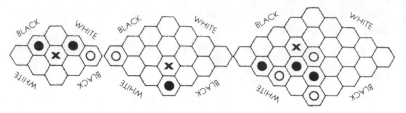

Figure 41. (Artist: Amy Kasai.)

ANSWERS

Solutions to the three Hex problems given in Figure 38 are shown in Figure 41. A complete analysis of alternate lines of play is too lengthy to give; only the one correct first move for White is indicated by the crosses.

Several readers expressed a belief that in the third problem White could also win by playing on cell 22 (begin at the extreme left and number the rows up and to the right from 1 to 25). Black, however, can defeat this by the following ingenious line of play:

White	Black
22	19
18	10
5	9
4	8
3	7

White's moves are forced in the sense that Black has a quicker win unless White makes the indicated move. At the close of the above moves, Black will have a chain with a double play at both ends and a double play to close one break within the chain, so there is no way White can prevent the win.

AFTERWORD

Many of my later *Scientific American* columns reported on Piet Hein's other creations. See Book 2 and Book 11 in my series for chapters on Piet Hein's Soma Cube, and Book 6 for a chapter on his superellipse and supereggs. Many of his books of delightful light verse, which he calls "grooks," have been translated and published in the United States.

Several topological games have been invented that are like Hex in that the winner is the first to complete a specified path joining two sides of a board. One of them, sold under the name of Bridg-it, turned out to have a winning line of play based on a pairing strategy (see Chapter 18 of Book 3). In contrast, Hex has proved unusually impervious to analysis. To this day, winning lines of play are known only for low-order boards, with no sign of a general strategy for boards of all sizes.

Rex, a name for the reverse form of Hex in which the winner forces his opponent to complete a chain, is (like the reverse forms of most two-person games) much harder to analyze. Ronald Evans (see his 1974 paper) has carried the analysis of Rex a step further than Robert Winder did by showing that on even-order boards White has a win by opening in an acute-corner cell. This is easily demonstrated on the 2×2 board, and it is easy to see the win on the 3×3, but the 4×4 is so complex that a winning line of play for the first player remains unknown. David Silverman reported in a letter that he had found an unusual pairing strategy for a second-player win on the 5×5. Readers will find a good 4×4 Rex problem by Evans in Book 12.

"Beck's Hex," a variant proposed by Anatole Beck, allows Black to tell White where he must make his first move. Beck was able to prove (see bibliography) that Black can always win by telling White to open in an acute-corner cell. Other variants of Hex, including Vex, Vertical Vex, Reverse Vertical Vex, and Tex are described in Evans's 1975 article.

Although it is intuitively apparent that Hex can't end in a draw, a formal proof is trickier than you might think. One "proof," published several times, goes like this: Imagine a finished game of Hex played on a paper diamond. Cut out all the cells on which Black played; then seize White's edges and pull. If the paper comes apart it shows that Black has completed a path. If it doesn't come apart, White has completed a path, and the paper is sure to come apart if you pull on Black's edges. Because one of the two must occur, a player must win. Unfortunately, it's not clear from what has been said that one of the two must occur. To assume that the diamond must pull apart in one direction or the other is to assume what one is seeking to prove, namely that one side must have completed a path. It remains to be shown that a game cannot end with all cells filled and neither side having completed a chain. As David Gale points out in his 1979 paper, the situation is similar to topology's notorious Jordan curve theorem. It is easy to "see" that a simple closed curve must cut the plane into two separate regions, but a formal proof is not so easy to formulate.

POSTSCRIPT

John Nash's mental problems and his brilliant mathematical achievements (they won him a Nobel Prize), are detailed in *A Beautiful Mind* and its movie version, where Russell Crowe plays John Nash. I had the pleasure of meeting Nash briefly when he was a graduate student at New York University and I was researching my *Scientific American* column on Hex.

Parker Brothers made and sold a boxed Hex set complete with a folded board and counters of two different colors. It came close to being featured as a game played on a spaceship in Arthur Clarke's movie *2001*, but at the last moment it was abandoned for a more picturesque three-dimensional chess board.

The proof that the first player can always win in Hex applies also to ticktacktoe games. If the game can end in a draw, the first player can always win or force a draw.

BIBLIOGRAPHY

"Computers and Automata." Claude Shannon in *Proceedings of the Institute of Radio Engineers* 41 (October 1953): 1234ff.

Signs, Symbols, and Noise. John Pierce. *Harper's* (1961): 10–13.

"The Game of Hex." Anatole Beck in *Excursions in Mathematics*, eds. Anatole Beck, Michael Bleicher, and Donald Crowe. Worth, 1969.

"A Winning Opening in Reverse Hex." Ronald Evans in *Journal of Recreational Mathematics* 7 (Summer 1974): 189–192.

"Some Variants of Hex." Ronald Evans in *Journal of Recreational Mathematics* 8 (1975–1976): 120–122.

"Hex Must Have a Winner." David Berman in *Mathematics Magazine* 49 (March 1976): 85–86. See two commenting letters in the same volume, page 156.

"The Game of Hex and the Brouwer Fixed Point Theorem." David Gale in *American Mathematical Monthly* 86 (December 1979): 818–827.

"Some Remarks about a Hex Problem." Claude Berge in *The Mathematical Gardner*, ed. David Klarner. Prindle, 1981, 25–27.

"Kriegspiel Hex." Duane Broline, solution to Problem 1091, in *Mathematics Magazine* 84 (March 1981): 85–86.

"Dodgem and Other Simple Games." Martin Gardner in *Time Travel and Other Mathematical Bewilderments*. Chapter 12. W. H. Freeman, 1988.

"Hex Games and Twixt Maps on the Annulus." S. Alpern and Anatole Beck in *American Mathematical Monthly* 98 (1991): 803–811.

"Who Wins Misère Hex?" Jeffrey Legarias and Donald Sleator in *The Mathematician and Pied Puzzler*, eds. Elwyn Berlekamp and Tom Rodgers. A K Peters, 1999, pp. 237–249.

"Hex Marks the Spot." Ian Stewart in *Scientific American* 283 (September 2000): 100–103.

"Three Hex Variants." Anatole Beck in *Abstract Games* 5 (2001): 34–35.

"Hex Strategy: Computer Hex." Cameron Browne in *Abstract Games* 8 (2001): 17–21.

"A Hierarchical Approach to Computer Hex." V. Anshelevich in *Artificial Intelligence* 134 (2002): 101–120.

"The Game of Hex." J. Milnor in *The Essential John Nash*, H. W. Kuhn and S. Nasar. Princeton University Press, 2002, pp. 29–33.

Connection Games: Variations on a Theme. Cameron Browne. A K Peters, 2005. A connection game is a board game in which players take turns placing a piece so as to make a specified path between two cells. This amazing book examines more than 200 such games!

"Random Turn Hex and Other Connection Games." Yuval Peres et al. in *American Mathematical Monthly* 114 (May 2007): 373–387.

CHAPTER NINE

Sam Loyd: America's Greatest Puzzlist

THE NAME SAM LOYD will not be familiar to many readers of this book, yet Loyd was an authentic American genius, and in his time something of a celebrity. For almost half a century, until his death in 1911, he was the nation's undisputed puzzle king. Thousands of superb puzzles, most of them mathematical, appeared under his name; many are still popular today.

Actually there were two Sam Loyds – father and son. When the elder Loyd died, the younger dropped the "Jr." from his name and continued his father's work, writing puzzle columns for magazines and newspapers and issuing books and novelties from a dingy little office in Brooklyn. But the son, who died in 1934, did not possess the father's inventiveness; his books are little more than hastily assembled compilations of his father's work.

Loyd senior was born in Philadelphia in 1841 of (as he once put it) "wealthy but honest parents." In 1844 his father, a real estate operator, moved his family to New York, where Sam attended public school until he was 17. If he had gone to college he might well have become an outstanding mathematician or engineer, but Sam did not go to college. One reason was that he had learned to play chess.

For 10 years, Loyd apparently did little except push chess pieces about on a chessboard. At that time, chess was enormously popular; many newspapers carried chess columns featuring problems devised by readers. Loyd's first problem was published by a New York paper when he was 14. During the next five years, his output of chess puzzles was so prodigious that he became known throughout the chess world. When he was 16, he was made problem editor of *Chess Monthly*, at that time edited by D. W. Fiske and the young

Figure 42. (Artist: Beverly Ruedi.)

chess master, Paul Morphy. Later he edited several newspaper chess columns and contributed regularly, under various pseudonyms, to a score of others.

In 1877 and 1878, Loyd wrote a weekly chess page for *Scientific American Supplement*, beginning each article with an initial letter formed by the pieces of a chess problem. These columns comprised most of his book *Chess Strategy*, which he printed in 1878 on his own press in Elizabeth, New Jersey. Containing 500 of his choicest problems, this book is now much sought by collectors.

Loyd's most widely reprinted chess problem, composed when he was 18, illustrates the delightful way in which his posers were often dressed up with anecdotes. It seems that in 1713, when Charles XII of Sweden was besieged by the Turks at his camp in Bender, the king often passed the time by playing chess with one of his ministers. On one occasion, when the game reached the situation depicted in Figure 42, Charles (playing white) announced a checkmate in three moves. At that instant a bullet shattered the white knight. Charles studied the board again, smiled, and said he did not need the knight because he still had a mate in four moves. No sooner had he said this than a second bullet removed his pawn at king's rook 2. Unperturbed, Charles considered his position carefully and announced mate in five.

Figure 43. (Artist: Amy Kasai.)

The story has a topper. Years later a German chess expert pointed out that if the first bullet had destroyed the white rook instead of the knight, Charles still would have had a mate in six. Chess-playing readers may enjoy tackling this remarkable four-part problem.

The original version of Loyd's first commercially successful puzzle, drawn by himself in his late teens, is depicted in Figure 43. When the puzzle was cut along the dotted lines, its three rectangles could be arranged (without folding) so that the two jockeys rode the two donkeys. P. T. Barnum bought millions of these puzzles from Loyd and distributed them as "P. T. Barnum's Trick Donkeys." It is said that the puzzle earned young Loyd $10,000 in a few weeks; it is popular to this day.

From the mathematical standpoint, Loyd's most interesting creation is the famous "14-15" or "Boss" puzzle. It had a surprising revival in the late 1940s and can still be bought in novelty stores.

1	2	3	4
5	6	7	8
9	10	11	12
13	15	14	

Figure 44. (Artist: Amy Kasai.)

As shown in Figure 44, 15 numbered squares are free to slide about within a box. At the beginning of the puzzle, the last two numbers are not in serial order. The problem is to slide the squares, without lifting them from the box, until all of them are in serial order, with the vacant space in the lower right-hand corner as before. In the 1880s the 14-15 puzzle had a tremendous vogue both here and abroad, and numerous learned articles about it appeared in mathematical journals.

Loyd offered a prize of $1,000 for a correct solution to the puzzle. Thousands of people swore they had solved it, but no one could recall his moves well enough to record them and collect the prize. Loyd's offer was safe because the problem is not solvable. Of the more than 20 trillion possible arrangements of the squares, exactly half can be made by sliding the squares from the arrangement depicted here. The remaining positions, including the one sought, have a different *parity* (to use the language of permutation mathematics) and cannot be reached from any position possessing the opposite parity.

The game was sometimes played by placing the squares in the box at random, and then trying to slide them into serial order. The probability of succeeding is of course 1/2. A simple way to determine whether any arrangement B can be obtained from any arrangement A is to see how many *interchanges* (exchanging the positions of any two squares by removing them from the box and replacing them) are required to convert A to B. If this number is even, A and B have the same parity and either can be obtained from the other by sliding.

The fact that a single interchange of any two blocks automatically reverses the parity underlies a particularly fiendish version of the

Figure 45. (Artist: Amy Kasai.)

puzzle marketed a few years ago. Here the squares are not numbered but lettered as shown in Figure 45: RATE and YOUR are on squares of one color, and MIND and PAL are on squares of another color. You show this arrangement to your victim and then destroy it by sliding the blocks here and there at random. As you do so you slyly maneuver the second R into the upper left-hand corner before you hand over the puzzle. The victim naturally permits this R to stay in the corner while he tries to put the rest of the blocks in order – an impossible feat because the switch of R's has switched the parity. The best the poor fellow can achieve is RATE YOUR MIND PLA.

Loyd's greatest puzzle is unquestionably his famous "Get off the Earth" paradox which he patented in 1896. A cardboard circle, riveted at the center to a square piece of cardboard, bears around its rim the pictures of 13 Chinese warriors. Part of each warrior is on the circle, and part on the square. When the wheel is turned slightly, the parts fit differently, and one warrior completely disappears! This puzzle has been reproduced so often that we show in Figure 46 the less familiar, but in some ways more puzzling, version called "Teddy and the Lions." In one position of the wheel, you see seven lions and seven hunters; in another, eight lions and six hunters. Where does the eighth lion come from? Which hunter vanishes and where does he go?

In 1914, three years after his father's death, Loyd junior issued a mammoth *Cyclopedia of Puzzles*, surely the greatest collection of problems ever assembled in one volume. The following brain teaser is taken from this fabulous, long-out-of-print work. It illustrates how cleverly the old master was able to take a simple problem, calling for nothing more than the ability to think clearly and handle fractions, and dramatize it in such a way that it becomes an exciting challenge.

Figure 46. Loyd's "Teddy and the Lions" paradox. On the top there are seven lions and seven hunters; on the bottom, eight lions and six hunters.

In Siam, Loyd explains, two kinds of fish are raised for their fighting qualities – a large white perch known as the kingfish and a small black carp called the devilfish. "Such antipathy exists between these two species that they attack each other on sight and battle to the death."

A kingfish can easily dispose of one or two of the little fish in just a few seconds. But the devilfish "are so agile and work together so harmoniously that three of the little fellows would just equal a big one, and they would battle for hours without results. So cleverly and scientifically do they carry on their line of attack that four of the little fellows would kill a large one in just three minutes and larger numbers would administer the *coup de grâce* proportionately quicker."

(That is, five devilfish would kill one kingfish in 2 minutes and 24 seconds, six in 2 minutes, and so on.)

If 4 kingfish are opposed to 13 devilfish, which side will win the fight and exactly how long will it take, assuming of course that the little fish cooperate in the most efficient manner?

To avoid an ambiguity in Loyd's statement of the problem, it should be made clear that the devilfish always attack single kingfish in groups of three or more, and stay with the large fish until he is disposed of. We cannot, for example, assume that while the 12 little fish hold the 4 large fish at bay, the thirteenth devilfish darts back and forth to finish off the large fish by attacking all of them simultaneously. If we permit fractions, so to speak, of devilfish to be effective, then we can reason that if 4 devils kill a king in 3 minutes, 13 devils will finish a king in 12/13 minutes, or 4 kings in 48/13 minutes (3 minutes, 41 and 7/13 seconds). But this same line of reasoning would lead to the conclusion that 12 devils would kill 1 king in 1 minute, or 4 kings in 4 minutes, even without the aid of the thirteenth little fish – a conclusion that clearly violates Loyd's assumption that three little fish are unable to kill one devil.

ADDENDUM

Arthur W. Burks, professor of philosophy at the University of Michigan, wrote to tell me of the interesting way in which Loyd's 14-15 puzzle resembles a modern digital computer. Each has a finite number of states, each state followed by another state. On every "run" of the computer or 14-15 puzzle, it begins in a certain state. All other states can then be divided into two groups: the "admissible" states, which can be realized by "inputs," and the "inadmissible" states, which cannot. The matter is discussed on page 63 of "The Logic of Fixed and Growing Automata" by Professor Burks; a 1957 memo issued by the Engineering Research Institute of the University of Michigan.

ANSWERS

In the chess problem, White mates in three by taking the pawn with his rook. If black bishop takes rook, White jumps his knight to B3, Black is forced to move his bishop and White mates with pawn to Kt4. If Black had taken the knight instead of the rook, white rook checks on R3, Black interposes bishop, White mates with pawn to Kt4 as before.

After the bullet shatters the white knight, White mates in four by taking the pawn with his pawn. If Black moves bishop to K6, White moves rook to Kt4. Black bishop to Kt4 is followed by white rook to R4 (check). Bishop takes rook and White mates with pawn to Kt4.

After the bullet removes the white pawn at R2, White mates in five with rook to QKt7. Should Black move his bishop to K6, then: (2) R-Kt1, B-Kt4; (3) R-KR1 (check), B-R5; (4) R-R2, PxR; (5) P-Kt4 (mate). Should Black on his first move play B-Kt8, then: (2) R-Kt1, B-R7; (3) R-K1, K-R5; (4) K-Kt6, any move; (5) R-K4 (mate).

If the first bullet had removed White's rook instead of his knight, White mates in six by moving knight to B3. Black's best response is B-K8, which leads to (2) KtxB, K-R5; (3) P-R3, K-R4; (4) Kt-Q3, K-R5; (5) Kt-B4, P-R4; (6) Kt-Kt6 (mate).

The jockeys can be placed on the two donkeys (which miraculously break into a gallop) as shown in Figure 47. Figure 48 reproduces a possible source of Loyd's famous puzzle: a Persian design of the early seventeenth century.

Figure 49 shows an old painting on a leather pillow that was originally owned by Maurits Escher. It was given to me after Escher's death by his eldest son George. Note that each horse's head belongs to *three* bodies. I in turn gave the painting to Stephen Turner, an

Figure 47. The puzzle of the donkeys solved. (Artist: Amy Kasai.)

Escher collector in Winston Salem, NC. The artist and date of the painting are unknown.

Concerning the "Teddy and the Lions" paradox, it is meaningless to ask which lion has vanished or which hunter has newly appeared.

Figure 48. Seventeenth-century Persian design. (Courtesy, Museum of Fine Arts, Boston.)

Figure 49. Old painting originally owned by Maurits Escher.

All the lions and hunters vanish when the parts are rearranged – to form a new set of eight lions, each 1/8 smaller than before, and six hunters, each 1/6 larger than before.

There are many ways to tackle the fighting-fish problem. Here is Loyd's own characteristic account of the solution:

> Three of the little fish paired off with each of three big fish, engaging their attention while the other four little fighters polished off the fourth big one in just three minutes. Then five little fellows tackled one big fish and killed him in 2 minutes 24 seconds; while the other little ones were battling with the other big ones.
>
> It is evident that if the remaining two groups had been assisted by one more fighter they would all have finished in the same time, so there is only sufficient resistance left in each of the big ones to call for the attention of a little fish for 2 minutes 24 seconds. Therefore if seven now attack instead of one, they would do it in one seventh of that time, or 20 and 4/7 seconds.
>
> In dividing the little-fish forces against the remaining two big ones – one would be attacked by seven and the other by six – the last fish at the end of the 20 and 4/7 seconds would still require the

punishment which one little one could administer in that time. The whole 13 little fellows, concentrating their attack, would give the fish his quietus in one thirteenth of that time, or 1 and 53/91 seconds.

Adding up the totals of the time given in the several rounds – 3 minutes, 2 minutes 24 seconds, 20 and 4/7 seconds, and 1 and 53/91 seconds, we have 5 minutes 46 and 2/13 seconds as the entire time consumed in the battle.

AFTERWORD, 1988

For more about Sam Loyd, readers are referred to the introductions to the two Dover collections of Loyd puzzles I have edited and to the chapter on advertising premiums in my Book 10. I was long under the impression that the collection of puzzles in Loyd's *Cyclopedia* had been selected by Loyd junior after his father's death. Not so. In 1907 Loyd senior started *Our Puzzle Magazine* (I do not know how many issues appeared), which drew upon his earlier output. After his father's death, Sam Loyd, Jr., simply printed the *Cyclopedia* from the plates of this periodical. Massive as the book is, it fails to include hundreds of Loyd puzzles that appeared in various publications and as advertising devices. Will Shortz, a former editor of *Games* magazine and now crossword puzzle editor of *The New York Times*, has for years been tracking down this rich mine of neglected material. I hope that he will some day edit a definitive collection of Loyd's forgotten creations.

Loyd's famous "Get off the Earth" premium was based on earlier linear versions. For a history of such paradoxes, see the two chapters on geometrical vanishes in my *Mathematics, Magic, and Mystery*, and the article by Mel Stover. Both references also cover related vanishing area paradoxes such as Paul Curry's triangle, presented here in Chapter 14. In recent years, the most popular version of a vanishing person is the disappearing leprechaun paradox, drawn by Ms. Pat Patterson, a Canadian artist. I reproduce and discuss it in the chapter on advertising premiums in my Book 10 and in *Aha, Gotcha!*

Loyd's 14-15 puzzle is covered in most of the classic general works on recreational mathematics: Rouse Ball in England, Edouard Lucas in France, W. Ahrens in Germany, and many others. In the second volume of his *Mathematische Unterhaltengen und Spiele* (Leipzig,

Figure 50. Illustration from Sam Loyd's *Cyclopedia of Puzzles.*

1918), Ahrens has an extensive treatment of the puzzle and its history. An eight-block simplification (3 × 3 matrix) is offered as a problem in Chapter 20 of my Book 5. Even simpler versions, such as five blocks on a 2 × 3 field, have been the basis of problems in many old puzzle books.

Unit squares provide the simplest kind of sliding-block puzzles. Hundreds of more difficult sliding puzzles – pieces with rectangular and other shapes, and fields with shapes other than square or rectangular – have been marketed from time to time around the globe. Chapter 7 of my Book 5 introduces this large field of mechanical puzzles. The definitive book about them, with numerous color photographs, is by L. E. Hordern (see bibliography, *Sliding Piece Puzzles*).

POSTSCRIPT

On previous pages I falsely credited Sam Loyd with having invented the 14-15 puzzle. I did so because Loyd himself took credit for this puzzle. (Figure 50 is an illustration of the puzzle from his 1914 book.) It turns out that he lied. Not only did he *not* invent the puzzle, he had almost nothing to do with its amazing popularity! For a detailed history of the puzzle, which became a mania in the 1880s throughout America and Europe, see Jerry Slocum's marvelous book *The 15 Puzzle: How it Drove the World Crazy*, by Slocum and Dic Sonneveld (published by Slocum in 2006). The craze even surpassed the recent craze of Rubik's cube!

Slocum's beautiful book is packed with photos of versions of the puzzle that were marketed here and abroad, along with cartoons about the mania, and comic poems telling how persons were driven out of their wits by fruitless efforts to solve the puzzle. Loyd not only claimed he invented the puzzle, but he also shamelessly took credit for the board game of Parchesi and for a skill puzzle called Pigs in Clover. Moreover, many of his published puzzles, in newspapers and magazines, were cribbed without credit from the writings of England's great puzzle maker Henry Dudeney.

Loyd does deserve credit for his vanishing Chinese warrior paradox (its title "Get off the Earth" reflected American prejudice against the flood of Chinese entering the country as laborers), and several clever variants that Loyd marketed as advertising premiums. The puzzle with the two donkeys was also his. For Mel Stover's hilarious switch on the donkeys see "Just for the Mel of It," by Max Maven in *Puzzler's Tribute*, David Wolfe and Tom Rodgers (eds.), A K Peters, 2002, pp. 21–27.

For 3 × 3 versions of the 14-15 puzzle, with minimum-move solutions, see my Book 5, Chapter 20.

Loyd's "Get off the Earth" paradox was a circular form of earlier linear versions of vanishing objects. See the two chapters on geometrical vanishes in my 1956 Dover paperback, *Mathematics, Magic, and Mystery*, and Mel Stover's *Games* article listed in the bibliography.

BIBLIOGRAPHY

BY LOYD

Chess Strategy: A Treatise on the Art of Problem Composition. Sam Loyd. Elizabeth, New Jersey: privately printed, 1878. A collection of about 500 chess problems by Loyd.
Sam Loyd's Puzzles: A Book for Children. Eds. Sam Loyd, Jr., and David McKay, 1912.

Sam Loyd's Cyclopedia of 5000 Puzzles, Tricks, and Conundrums. Ed.
Sam Loyd, Jr. Lamb Publishing Co., 1914.
Sam Loyd's Picture Puzzles. Ed. Sam Loyd, Jr. New York: privately
printed, 1924.
Sam Loyd and His Puzzles. Ed. Sam Loyd, Jr. Barse, 1928.
Mathematical Puzzles of Sam Loyd. Ed. Martin Gardner, Vols. 1 and 2.
Dover, 1959 and 1960.
Eighth Book of Tan: Seven Hundred Tangrams. Sam Loyd. Dover, 1968.
Reprint of one of Loyd's rarest books.

ABOUT LOYD

"The Prince of Puzzle-Makers." Interview with Loyd by George Bain in
Strand Magazine 34 (December 1907): 771–777.
"My Fifty Years in Puzzleland." Interview with Loyd by Walter Eaton in
Delineator (April 1911): 274ff.
Sam Loyd and His Chess Problems. Alain White. Whitehead and Miller,
1913. Dover reprint, 1962.
"Advertising Premiums." Martin Gardner in *Wheels, Life, and Other
Mathematical Amusements.* Chapter 12. W. H. Freeman, 1983.

THE FIFTEEN PUZZLE

"Notes on the '15' Puzzle." William Johnson in *American Journal of
Mathematics* 2 (1879): 397–399.
"Notes on the '15' Puzzle, II." William Story in *American Journal of
Mathematics* 2 (1879): 399–404.
"A New Look at the Fifteen Puzzle." E. E. Spitznagel, Jr., in *Mathematics
Magazine* 40 (September 1967): 171–174.
"Rotating the Fifteen Puzzle." A. L. Davies in *Mathematical Gazette* 54
(October 1970): 237–240.
"Some Generalizations of the 14-15 Puzzle." Hans Liebeck in *Mathe-
matics Magazine* 44 (September 1971): 185–189.
"Systematic Solutions of the Famous 14-15 Puzzles." Alan Henney and
Dagmar Henney in *Pi Mu Epsilon Journal* 6 (Spring 1976): 197–201.
"The 14-15 Puzzle." A. K. Austin in *Mathematical Gazette* 63 (March
1979): 45–46.
Sliding Piece Puzzles. L. E. Horden. Oxford University Press, 1986. The
bibliography lists some fifty references on the 14-15 puzzle.
"A Modern Treatment of the 15 Puzzle." A. F. Archer. *American Mathe-
matical Monthly* 106 (1999): 793–799.

The 15 Puzzle. Jerry Slocum and Dic Sonneveld. Slocum Puzzle Foundation, 2006.

"The 15 Puzzle." Jerry Slocum and Dic Sonneveld. Reviewed by Aaron Archer in *The Mathematical Intelligencer* 29 (2007): 83–85.

GEOMETRICAL VANISHES

Mathematics, Magic, and Mystery. Martin Gardner. Chapters 7 and 8. Dover, 1956.

"The Disappearing Man and Other Vanishing Paradoxes." Mel Stover in *Games* (November–December 1980): 14–18.

CHAPTER TEN

Mathematical Card Tricks

SOMERSET MAUGHAM's short story "Mr. Know-All" contains the following dialogue:

"Do you like card tricks?"
"No, I hate card tricks."
"Well, I'll just show you this one."

After the third trick, the victim finds an excuse to leave the room. His reaction is understandable. Most card magic is a crashing bore unless it is performed by skillful professionals. There are, however, some "self-working" card tricks that are interesting from a mathematical standpoint.

Consider the following trick. The magician, who is seated at a table directly opposite a spectator, first reverses 20 cards anywhere in the deck. That is, he turns them face up in the pack. The spectator thoroughly shuffles the deck so that these reversed cards are randomly distributed. He then holds the deck underneath the table, where it is out of sight of everyone, and counts off 20 cards from the top. This packet of 20 cards is handed under the table to the magician.

The magician takes the packet but continues to hold it beneath the table so that he cannot see the cards. "Neither you nor I," he says, "knows how many cards are reversed in this group of 20 that you handed me. However, it is likely that the number of such cards is less than the number of reversed cards among the 32 that you are holding. Without looking at my cards, I am going to turn a few more face-down cards face up and attempt to bring the number of reversed cards in my packet to exactly the same number as the number of reversed cards in yours."

The magician fumbles with his cards for a moment, pretending that he can distinguish the fronts and backs of the cards by feeling them. Then he brings the packet into view and spreads it on the table. The face-up cards are counted. Their number proves to be identical with the number of face-up cards among the 32 held by the spectator!

This remarkable trick can best be explained by reference to one of the oldest mathematical brain-teasers. Imagine that you have before you two beakers, one containing a liter of water; the other a liter of wine. One cubic centimeter of water is transferred to the beaker of wine and the wine and water mixed thoroughly. Then a cubic centimeter of the mixture is transferred to the water. Is there now more water in the wine than wine in the water? Or vice versa? (We ignore the fact that in practice, a mixture of water and alcohol is a trifle less than the sum of the volumes of the two liquids before they are mixed.)

The answer is that there is just as much wine in the water as water in the wine. The amusing thing about this problem is the extraordinary amount of irrelevant information involved. It is not necessary to know how much liquid there is in each beaker, how much is transferred, or how many transfers are made. It does not matter whether the mixtures are thoroughly mixed or not. It is not even essential that the two vessels hold equal amounts of liquid at the start! The only significant condition is that at the end each beaker must hold exactly as much liquid as it did at the beginning. When this obtains, then obviously if x amount of wine is missing from the wine beaker, the space previously occupied by the wine must now be filled with x amount of water.

If the reader is troubled by this reasoning, he can quickly clarify it with a deck of cards. Place 26 cards face down on the table to represent wine. Beside them put 26 cards face up to represent water. Now you may transfer cards back and forth in any manner you please from any part of one pile to any part of the other, provided you finish with exactly 26 in each pile. You will then find that the number of face-down cards in either pile will match the number of face-up cards in the other pile.

Now try a similar test beginning with 32 face-down cards and 20 face-up cards. Make as many transfers as you wish, ending with

20 cards in the smaller pile. The number of face-up cards in the large pile will of necessity exactly equal the number of face-down cards among the 20. Now turn over the small pile. This automatically turns its face-down cards face up and its face-up cards face down. The number of face-up cards in both groups will therefore be the same.

The operation of the trick should now be clear. At the beginning the magician reverses exactly 20 cards. Later, when he takes the packet of 20 cards from the spectator, it will contain a number of face-down cards equal to the number of face-up cards remaining in the deck. He then pretends to reverse some additional cards, but actually all he does is turn the packet over. It will then contain the same number of reversed cards as there are reversed cards in the group of 32 held by the spectator. The trick is particularly puzzling to mathematicians, who are apt to think of all sorts of complicated explanations.

Many card effects known in the conjuring trade as spellers are based on elementary mathematical principles. Here is one of the best. With your back to the audience, ask someone to take from 1 to 12 cards from the deck and hide them in his pocket without telling you the number. You then tell him to look at the card at that number from the top of the remainder of the deck and remember it.

Turn around and ask for the name of any individual, living or dead. For example, someone suggests Marilyn Monroe (the name, by the way, must have more than 12 letters). Taking the deck in your hand, you say to the person who pocketed the cards: "I want you to deal the cards one at a time on the table, spelling the name Marilyn Monroe like this." To demonstrate, deal the cards from the top of the deck to form a face-down pile on the table, taking one card for each letter until you have spelled the name aloud. Pick up the small pile and replace it on the deck.

"Before you do this, however," you continue, "I want you to add to the top of the deck the cards you have in your pocket." Emphasize the fact, which is true, that you have no way of knowing how many cards this will be. Yet in spite of this addition of an unknown number of cards, after the spectator has completed spelling Marilyn Monroe, the next card (that is, the card on top of the deck) will invariably turn out to be his chosen card!

The operation of the trick yields easily to analysis. Let x be the number of cards in the spectator's pocket and also the position of the chosen card from the top of the deck. Let y be the number of letters in the selected name. Your demonstration of how to spell the name automatically reverses the order of y cards, bringing the chosen card to a position from the top that is y minus x. Adding x cards to the deck therefore puts y minus x plus x cards above the selected one. The x's cancel out, leaving exactly y cards to be spelled before the desired card is reached.

A more subtle compensatory principle is involved in the following effect. A spectator is asked to select any three cards and place them face down on the table without letting the magician see them. The remaining cards are shuffled and handed to the magician.

"I will not alter the position of a single card," the magician explains. "All I shall do is remove one card that will match in value and color the card you will select in a moment." He then takes a single card from the pack and places it face down at one side of the table.

The spectator is now asked to take the remaining cards in hand and to turn face up the three cards he previously placed on the table. Let us assume that they are a nine, a queen, and an ace. The magician requests that he start dealing cards face down on top of the nine, counting aloud as he does so, beginning the count with "10" and continuing until he reaches "15." In other words, the spectator deals six cards face down on the nine. The same procedure is followed with the other two cards. The queen, which has a value of 12 (jacks are 11; kings, 13), will require three cards to bring the count from 12 to 15. The ace (1) will require 14 cards.

The magician now has the spectator total the values of the three original face-up cards, and note the card at that position from the top of the remainder of the deck. In this case the total is 22 (9 plus 12 plus 1), so he looks at the twenty-second card. The magician turns over his "prediction card." The two cards match in value and color!

How is it done? When the magician glances through the deck to find a "prediction card," he notes the fourth card from the bottom and then removes another card that matches it in value and color. The rest of the trick works automatically. (On rare occasions, you may find the prediction card among the bottom three cards of the pack. When this happens you must remember to tell the spectator later, when he makes his final count to a selected card, to finish the count, then look at the *next* card.) I leave to the reader the easy task of working out an algebraic proof of why the trick cannot fail.

The ease with which cards can be shuffled makes them peculiarly appropriate for demonstrating a variety of probability theorems, many of which are startling enough to be called tricks. For example, let us imagine that two people each hold a shuffled deck of 52 cards. One person counts aloud from 1 to 52; on each count, both deal a card face up on the table. What is the probability that at some point during the deal two identical cards will be dealt simultaneously?

Most people would suppose the probability to be low, but actually it is better than $1/2$! The probability there will be *no* coincidence is 1 over the transcendental number e. (This is not precisely true, but the error is less than 1 divided by 10^{69}. The reader may consult page 47 in the current edition of W. Rouse Ball's *Mathematical Recreations and Essays* for a method of arriving at this figure.) Since e is $2.718\ldots$, the probability of a coincidence is roughly $17/27$ or almost $2/3$. If you can find someone willing to bet you even odds that no coincidence will occur, you stand a rather good chance to pick up some extra change. It is interesting to note that we have here an empirical procedure, based on probability, for making a decimal expansion of e (comparable to the Buffon's Needle procedure for doing the same thing with pi). The more cards used, the closer the probability of no coincidence approaches $1/e$.

POSTSCRIPT

Since Dover published my *Mathematics, Magic, and Mystery* in 1956, the literature on tricks based on mathematical principles has grown exponentially. Thousands of clever card effects that are what magicians call "self-working" have been published in books and magic magazines. Living magic buffs who have been especially creative in this curious field are mathematician Colm Mulcahy, of Spelman College; Nick Trost, writer of several books on self-working card magic; and Jim Steinmeyer, creator of stage illusions and author of books about magic and magicians. Some of Steinmeyer's card effects, including a famous spelling trick with nine cards, are in his small books titled *Impuzzibilities* and *Further Impuzzibilities*. Books by Karl Fulves and Harry Lorrayne are also rich sources of mathematical card tricks.

Among magicians no longer living who were especially creative in the field of mathematical card tricks, one thinks of Charles Jordan, Alex Elmsley, and Bob Hummer. You'll find a raft of mathematical tricks with cards, coins, dice, and other objects in the 14 other books in this series.

BIBLIOGRAPHY

Chance and Choice by Cardpack and Chessboard: An Introduction to Probability in Practice by Visual Aids. Lancelot Hogben. Vol. 1. Chanticleer, 1950.

Scarne on Card Tricks. John Scarne. Crown, 1950.

Mathematics, Magic and Mystery. Martin Gardner. Dover, 1956.

Mathematical Magic. William Simon. Scribner's, 1964.

Self-Working Card Tricks. Karl Fulves. Dover, 1976.

CHAPTER ELEVEN

Memorizing Numbers

EVERYONE USES *mnemonic devices* – ways of memorizing bits of information by associating them with things that are easier to remember. In the United States, the most familiar of these devices is surely the rhyme beginning: "Thirty days hath September...." Another well-known mnemonic device is: "Every good boy does fine" (for EGBDF, the lines of the musical staff).

The same principle can also be applied, with ingenious variations, to the memorizing of numbers. Such tricks come easily to mathematicians. When Bertrand Russell visited New York in 1951, he told a newspaper columnist that he had no difficulty in recalling the number of his room at the Waldorf-Astoria – 1414 – because 1.414 is the square root of 2. The British mathematician G. H. Hardy wrote of calling on his friend Srinivasa Ramanujan, the Indian mathematical genius, in a taxicab numbered 1729. Hardy remarked that this was a dull number. "No," Ramanujan promptly replied. "It is a very interesting number. It is the smallest number expressible as a sum of two cubes in two different ways" (12 cubed plus 1 cubed, or 10 cubed plus 9 cubed). It must be admitted that even among mathematicians such an intimate acquaintance with numbers is rare.

The most common mnemonic device for remembering a series of digits is a sentence or rhyme in which the number of letters in each word corresponds to the digits in the desired order. Many such memory props have been worked out in various languages to recall pi beyond the usual four decimals. In English they range in length from the anonymous "May I have a large container of coffee?" through Sir James Jeans's "How I want a drink, alcoholic of course, after the heavy chapters involving quantum mechanics" to this doggerel contributed by Edouard Prevost to a publication of Graham Transmissions, Inc.:

> Now I – even I – would celebrate
> In rhymes inept the great
> Immortal Syracusan rivaled nevermore,
> Who by his wondrous lore,
> Untold us before,
> Made the way straight
> How to circles mensurate.

I know of no similar aids in English to recall *e*, the other common transcendental number. However, if you memorize *e* to five decimal places (2.71828), you automatically know it to nine, because the last four digits obligingly repeat themselves (2.718281828). In France *e* is memorized to 10 places by the traditional memory aid: *Tu aideras à rappeler ta quantité à beaucoup de docteurs amis*. Perhaps some reader can construct an amusing English sentence that will carry *e* to at least 20 decimals.

Is there a mnemonic system that, once it has been mastered, will enable one to memorize quickly any series of digits? There is such a system, and it has been developed to a high degree by modern memory experts. Not only can the system be used to give an impressive dinner-table demonstration of memory; it also can be highly useful in memorizing important mathematical and physical constants, historical dates, house and telephone numbers, license plates, social security numbers, and so on.

Although the art of mnemonics goes back to ancient Greece (the term comes from *Mnemosyne*, the Greek goddess of memory), it was not until 1634 that a Frenchman named Pierre Hérigone published in Paris his *Cursus Mathematici*, which contained an ingenious system for memorizing numbers. The system consisted in substituting consonants for digits and then adding vowels wherever required so that words would be formed. The words were then easily memorized by other mnemonic methods.

Hérigone's original number alphabet was soon adopted by memory experts in many countries. In Germany the great Gottfried Wilhelm von Leibniz was sufficiently intrigued by the notion to incorporate it into his scheme for a universal language; Lewis Carroll devised what he regarded as an improvement over the number alphabet in Richard Grey's *Memoria Technica*, a popular British work on mnemonics published in 1730. (A reproduction of Carroll's notes on his number alphabet will be found in Warren Weaver's article "Lewis Carroll: Mathematician," in *Scientific American* for April 1956.) In his diary, Carroll records that he applied his system to lines for memorizing pi to 71 decimals and to key words for the logarithms of all prime numbers less than 100. At one time, he planned to issue a booklet titled, *Logarithms by Lightning: A Mathematical Curiosity.*

The modern form of Hérigone's number alphabet, as currently used by all English-speaking memory experts, is shown in the chart of Figure 51. This must be thoroughly fixed in the memory before the system can be used profitably. On the right side of the chart are suggestions that may help in memorizing the table. Note that only consonants are employed, and that where two or more consonants stand for the same digit, they have similar sounds. Three consonants – W, H and Y (spelling "why") – do not appear on the chart.

Suppose we wish to use this system for remembering that mercury boils at 357 degrees centigrade. Our first step is to find a word in which the consonants, taken in order, will translate into 357. Such a word readily comes to mind – *MiLK*. The next step is to associate this word by a vivid mental picture with the word "mercury." One way to do this is to imagine Mercury, the messenger of the gods, winging his way through the clouds with a container of milk in his hands. The more preposterous the mental image the more easily it is retained by the mind. When we wish to recall the boiling point of mercury we have only to follow the chain of associations from the element to the Greek god to milk to 357. This may seem like a roundabout means of memorizing a number, but no better artificial system has yet been discovered. It is astonishing how firmly the links of the chain remain planted in the mind.

Consider some additional examples. The atomic number of the element indium is 49. We can recall this easily by linking India with the word *RuPee*. Neptunium has an atomic number of 93; we imagine Neptune puffing an *oPiuM* pipe. For tantalum, element 73, we might picture Tantalus plugging the hole in his tantalizing cup with

DIGITS	CONSONANTS	MEMORY AIDS	
1	**T, D**	**T** has one downstroke	**t**
2	**N**	**N** has two downstrokes	**n**
3	**M**	**M** has three downstrokes	**m**
4	**R**	**R** is the fourth letter in "four"	FOU**R**
5	**L**	**L** is 50 in Roman numerals	**50**
6	**J**, soft **G**, **SH**, **CH**	**J** looks like 6 when reversed	**J 6**
7	**K**, hard **G**, hard **C**	**K** can be printed with two sevens	𝒦
8	**F, V,** **PH** as in photo	**F** , in lower-case script, has two loops like the figure 8	ℓ **8**
9	**P, B**	**P** looks like 9 when reversed	**P 9**
0	**Z, S,** soft **C**	**Z** is the initial of "zero"	**Z**ERO

Figure 51. A "number alphabet" in which consonants stand for digits.

a wad of chewing *GuM*. Platinum, number 78, can be recalled by thinking of yourself sporting a pair of platinum *CuFF* links. Double letters, such as the f's in "cuff," are regarded as single letters. The number alphabet is strictly phonetic. Silent consonants, as well as W, H and Y, are ignored.

The chart of Figure 52 shows how the system can be used for memorizing to three decimal places the square roots of 2, 3, 5, 6, 7, 8, 10. (The square root of 8 is of course twice the square root of 2. Similarly, the square root of 12 can be obtained by doubling the square root of 3.) Only the first three consonants of each key word

NUMBER	SQUARE ROOT	MNEMONIC KEY
2	1.414	**R**AT **R**ACE. Think of two rats racing.
3	1.732	**KIMONO**. Three suggests triangle. Think of a kimono decorated with a pattern of small triangles.
5	2.236	**ENMESH**. Five suggests pentagon. Think of the pentagon hopelessly enmeshed in red tape.
6	2.449	**RARE BEE**. Six suggests hexagon. Think of the hexagonal cells of a beehive. Crawling over the cells is a two-headed bee.
7	2.645	**SHEER LINEN**. Seven suggests the dance of seven veils. Think of the veils as made of sheer linen.
8	2.828	**FUNNY FACE**. Eight suggests "ate." Think of taking a bite and making a funny face.
10	3.162	**TOUCH NOSE**. Ten suggests the fingers. Think of touching your nose with all ten of them.

Figure 52. How the number alphabet can be used to memorize square roots.

or phrase are considered. They stand for the three decimals of the corresponding square root. (The digit preceding the decimal point need not be considered because it is obvious.) Many other words can of course be substituted for those chosen here. It is usually best, in fact, to work out your own key words and mental associations rather than adopt those of someone else; your inventions will be closer to your own experience and therefore easier to recall.

Larger numbers can be memorized by taking figures in pairs or triplets, devising a suitable word for each group, and linking the words in a chain of striking mental pictures. A telephone number, for example, would be fixed in the memory by a chain of images connecting the person or firm to the exchange, then to two words that stand for the digits in the phone number.

It is by means of such chains of mental pictures that professional memory experts are able to repeat long lists of random digits immediately after the list has been read aloud to them. This seemingly incredible feat is well within the powers of anyone who troubles to spend a few weeks of daily practice in mastering the number alphabet. As a first step try memorizing the eight digits in the number on a dollar bill. Take the digits two at a time, forming words in which the first two consonants of each word correspond to a pair of numbers. For example, if the number is 41–09–15–85, these pairs can be translated into the four words: *ReD, ZeBra, TeLescope, FLower.* Think first of a *red zebra*. It holds a *telescope* to its eye. The telescope is trained on a distant *flower.*

In choosing words, nouns that provide vivid pictures are of course preferable, though adjectives can often be linked conveniently to a following noun, as in *red zebra.* In most cases the first words that come to mind are preferable, and each word should be linked to the next one by the most ridiculous image you can imagine. With practice, appropriate words will occur to you more rapidly, and you should soon be able to form your chain of mental pictures fast enough to keep pace with anyone who calls the digits to you slowly.

Memory experts are able to form chains of mental associations with extraordinary speed because every pair of digits immediately suggests to them a picture word taken from a previously memorized list. Thus they do not waste time in groping for suitable words. Some experts work with prememorized word lists for three-digit groups. To aid the students of his memory school in New York, Bruno Furst provides them with a printed number dictionary listing a variety of appropriate words for each number from 1 to 1,000. Such lists are not necessary, however, unless you intend to develop great proficiency in the art. Suitable words can always be devised as you go along if the numbers are read to you slowly, and you will discover that it is not at all difficult to memorize a series of 50 random digits by this method. Fortunately long chains of quickly improvised mental pictures do not remain long in the mind, so if you repeat the stunt a day or so later there will be no confusion of the new key words with those of the previous demonstration.

ADDENDUM

Among the many responses to my request for a mnemonic sentence for *e*, the following seemed to me particularly noteworthy:

> To express *e*, remember to memorize a sentence to simplify this. (John L. Greene, Beverly Hills, California.)
> To disrupt a playroom is commonly a practice of children. (Joseph J. Guiteras, Baldwinsville, New York.)
> By omnibus I traveled to Brooklyn. (David Mage, New York, New York.)
> It enables a numskull to memorize a quantity of numerals. (Gene Widhoff, Burbank, California.)

The *Enciclopedia universal ilustrada*, in an article on "Mnemotecnia," gives the following Spanish sentence for *e*: *Te ayudaré a recordar la cantidad a indoctos si reléesme bien.* Several Italian verses for *e* will be found on page 755 of *Matematica Dilettevole e Curiosa* by Italo Ghersi.

BIBLIOGRAPHY

"Mnemonic Verses and Words." Charles Peirce in *Baldwin's Dictionary of Philosophy and Psychology.* Macmillan, 1901.
"Mnemonics." John Malcolm Mitchell in *Encyclopedia Britannica.* 11th ed. 1911. Excellent history, with references to earlier books and bibliographies.
Memorizing Numbers. Bernard Zufall. New York: privately printed, 1940.
Stop Forgetting. Bruno Furst. Greenberg, 1949.
The Roth Memory Course. David Roth. This famous correspondence course of 1918 was issued as a single volume in 1954 by Writers Publishing Company, and later by Dial Press. It is still the best book on modern mnemonics.
How to Develop a Super-Powerful Memory. Harry Lorayne. Frederick Fell, 1957. Signet reprint, 1974.

Bibliography of Memory. Dr. Morris Young. Chilton, 1961. More than 6,000 references are cited in this bibliography by a Manhattan ophthalmologist and collector of books on memory systems.

Course in Memory and Concentration. Bruno Furst. A correspondence course that started in 1948. Later reprinted as ten booklets in a set entitled *You Can Remember!* Chicago: privately published, 1963.

Know Your Toes. William Jayme and Roderick Cook. Clarkson Potter, 1963.

The Art of Memory. Frances Yates. University of Chicago Press, 1966.

Mnemonics for Anatomy Students. David Gerrick. Dayton Labs, 1975.

Secrets of Mind Power. Harry Lorayne. Signet, 1975.

"Mnemonics." Martin Gardner in *Encyclopedia of Impromptu Magic.* Magic, Inc., 1978, 385–392.

Mnemonics, Rhetoric, and Poetics for Medics. Robert Bloomfield and Ted Chandler (vol. 1) and Carolyn Pedley et al. (vol. 2). Harbinger, 1982 and 1984.

Mnemonics and Tactics in Surgery and Medicine. John J. Shipman. Year Book Medical Publications, 1984.

The Memory Book. Jerry Lucas and Harry Lorayne. Ballantine, 1985.

The Absent-Minded Professor's Memory Book. Michele Slung. Ballantine, 1985.

Harry Lorayne's Page-a-Month Memory Book. Harry Lorayne. Ballantine, 1986.

CHAPTER TWELVE

Nine More Problems

1. THE TOUCHING CIGARETTES

Four golf balls can be placed so that each ball touches the other three. Five half-dollars can be arranged so that each coin touches the other four (see Figure 53).

Is it possible to place six cigarettes so that each touches the other five? The cigarettes must not be bent or broken.

2. TWO FERRYBOATS

Two ferryboats start at the same instant from opposite sides of a river, traveling across the water on routes at right angles to the shores. Each travels at a constant speed, but one is faster than the other. They pass at a point 720 yards from the nearest shore. Both boats remain in their slips for 10 minutes before starting back. On the return trips they meet 400 yards from the other shore.

How wide is the river?

Figure 53. (Artist: James Egleson.)

3. GUESS THE DIAGONAL

A rectangle is inscribed in the quadrant of a circle as shown (Figure 54). Given the unit distances indicated, can you accurately determine the length of the diagonal AC?

Time limit: one minute!

4. THE EFFICIENT ELECTRICIAN

An electrician is faced with this annoying dilemma. In the basement of a three-story house, he finds bunched together in a hole in the wall the exposed ends of 11 wires, all alike. In a hole in the wall on the top floor he finds the other ends of the same 11 wires, but he has no way of knowing which end above belongs to which end below. His problem: to match the ends.

To accomplish his task he can do two things: (1) short-circuit the wires at either spot by twisting ends together in any manner he wishes; (2) test for a closed circuit by means of a continuity tester consisting of a battery and a bell. The bell rings when the instrument is applied to two ends of a continuous, unbroken circuit.

Not wishing to exhaust himself by needless stair climbing, and having a passionate interest in operations research, the electrician sat down on the top floor with pencil and paper and soon devised the most efficient possible method of labeling the wires.

What was his method?

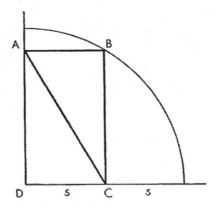

Figure 54. (Artist: James Egleson.)

Figure 55. (Artist: James Egleson.)

5. CROSS THE NETWORK

One of the oldest of topological puzzles, familiar to many a school-boy, consists of drawing a continuous line across the closed network shown in Figure 55 so that the line crosses each of the 16 segments of the network only once. The curved line shown here does not solve the puzzle because it leaves one segment uncrossed. No "trick" solutions are allowed, such as passing the line through a vertex or along one of the segments, folding the paper, and so on.

It is not difficult to prove that the puzzle cannot be solved on a plane surface. Two questions: Can it be solved on the surface of a sphere? On the surface of a torus (doughnut)?

6. THE TWELVE MATCHES

Assuming that a match is a unit of length, it is possible to place 12 matches on a plane in various ways to form polygons with integral areas. Figure 56 shows two such polygons: a square with an area of 9 square units, and a cross with an area of 5.

Figure 56. (Artist: James Egleson.)

Figure 57. (Artist: James Egleson.)

The problem is this: Use all 12 matches (the entire length of each match must be used) to form in similar fashion the perimeter of a polygon with an area of exactly 4 square units.

7. HOLE IN THE SPHERE

This incredible problem – incredible because it seems to lack sufficient data for a solution – appeared in a recent issue of *The Graham Dial*, a publication of Graham Transmissions, Inc. A cylindrical hole 6 inches long has been drilled straight through the center of a solid sphere. What is the volume remaining in the sphere?

8. THE AMOROUS BUGS

Four bugs – A, B, C, and D – occupy the corners of a square 10 inches on a side (Figure 57). A and C are male, B and D are female. Simultaneously A crawls directly toward B, B toward C, C toward D, and D toward A. If all four bugs crawl at the same constant rate, they will describe four congruent logarithmic spirals that meet at the center of the square.

How far does each bug travel before they meet? The problem can be solved without calculus.

9. HOW MANY CHILDREN?

"I hear some youngsters playing in the back yard," said Jones, a graduate student in mathematics. "Are they all yours?"

Figure 58. (Artist: James Egleson.)

"Heavens, no," exclaimed Professor Smith, the eminent number theorist. "My children are playing with friends from three other families in the neighborhood, although our family happens to be largest. The Browns have a smaller number of children, the Greens have a still smaller number, and the Blacks have the smallest of all."

"How many children are there altogether?" asked Jones.

"Let me put it this way," said Smith. "There are fewer than 18 children, and the product of the numbers in the four families happens to be my house number which you saw when you arrived."

Jones took a notebook and pencil from his pocket and started scribbling. A moment later he looked up and said, "I need more information. Is there more than one child in the Black family?"

As soon as Smith replies, Jones smiled and correctly stated the number of children in each family.

Knowing the house number and whether the Blacks had more than one child, Jones found the problem trivial. It is a remarkable fact, however, that the number of children in each family can be determined solely on the basis of the information given here!

ANSWERS

1. There are several different ways of placing the six cigarettes. Figure 58 shows the traditional solution as it is given in several old puzzle books.

    To my vast surprise, about 15 readers discovered that *seven* cigarettes could also be placed so that each touched all of the others! This of course makes the older puzzle obsolete. Figure 59, sent to me by George Rybicki and John Reynolds,

Figure 59. (Artist: James Egleson.)

graduate students in physics at Harvard, shows how it is done. "The diagram has been drawn," they write, "for the critical case where the ratio of length to diameter of the cigarettes is $^7/_2\sqrt{3}$. Here the points of contact occur right at the ends of the cigarettes. The solution obviously will work for any length-to-diameter ratio greater than $^7/_2\sqrt{3}$. Some observations on actual 'regular' size cigarettes give a ratio of about 8 to 1, which is, in fact, greater than $^7/_2\sqrt{3}$, so this is an acceptable solution." Note that if the center cigarette, pointing directly toward you in the diagram, is withdrawn, the remaining six provide a neat symmetrical solution of the original problem.

2. When the ferryboats meet for the first time (top of Figure 60), the combined distance traveled by the boats is equal to the

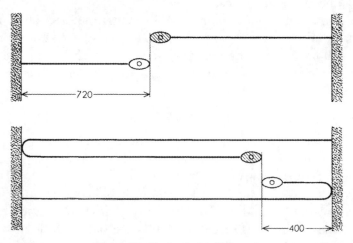

Figure 60. (Artist: James Egleson.)

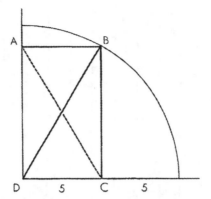

Figure 61. (Artist: James Egleson.)

width of the river. When they reach the opposite shore, the combined distance is twice the width of the river, and when they meet the second time (bottom of Figure 60), the total distance is three times the river's width. Since the boats have been moving at a constant speed for the same period of time, it follows that each boat has gone three times as far as when they first met and had traveled a combined distance of one river-width. Since the white boat had traveled 720 yards when the first meeting occurred, its total distance at the time of the second meeting must be 3 × 720, or 2,160 yards. The bottom illustration shows clearly that this distance is 400 yards more than the river's width, so we subtract 400 from 2,160 to obtain 1,760 yards, or one mile, as the width of the river. The time the boats remained at their landings does not enter into the problem.

The problem can be approached in other ways. Many readers solved it as follows. Let x equal the river-width. On the first trip the ratio of distances traveled by the two boats is $x - 720 : 720$. On the second trip it is $2x - 400 : x + 400$. These ratios are equal, so it is easy to solve for x. (The problem appears in Sam Loyd's *Cyclopedia of Puzzles*, 1914, page 80.)

3. Line AC is one diagonal of the rectangle (Figure 61). The other diagonal is clearly the 10-unit radius of the circle. Since the diagonals are equal, line AC is 10 units long.

4. On the top floor, the electrician shorted five pairs of wires (the shorted pairs are connected by broken lines in Figure 62),

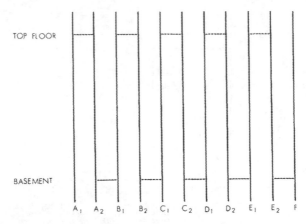

Figure 62. (Artist: James Egleson.)

leaving one free wire. Then he walked to the basement and identified the lower ends of the shorted pairs by means of his continuity tester. He labeled the ends as shown and then shorted them in the manner indicated by the dotted lines.

Back on the top floor, he removed all the shorts but left the wires twisted at insulated portions so that the pairs were still identifiable. He then checked for continuity between the free wire (which he knew to be the upper end of F) and some other wire. When he found the other wire, he was able at once to label it E2 and to identify its mate as E1. He next tested for continuity between E1 and another end which, when found, could be marked D2 and its mate D1. Continuing in this fashion, the remaining ends were easily identified. The procedure obviously works for any odd number of wires.

J. G. Fletcher, Princeton, New Jersey, was the first to send a method of applying this procedure, with a slight modification, to any even number of wires except two. Assume there is a twelfth wire on the far right in Figure 62. The same five pairs are shorted on the top floor, leaving two free wires. In the basement, the wires are shorted as before, and the twelfth wire is labeled G. Back on the top floor, G is easily identified as the only one of the two free wires in which no continuity is found. The remaining 11 wires are then labeled as previously explained.

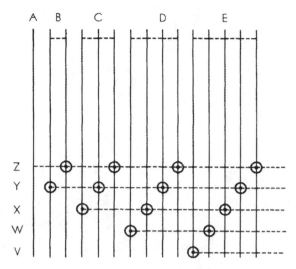

Figure 63. (Artist: James Egleson.)

In some ways a more efficient procedure, which takes care of all cases except two wires (two wires have no solution), was sent in by D. N. Buell, Detroit; R. Elsdon-Dew, Durban, South Africa; Louis Katz and Fremont Reizman, physics students at the University of Wisconsin; and Danforth K. Gannett, Denville, New Jersey. Mr. Gannett explained it clearly with the diagram for 15 wires shown in Figure 63. The method of labeling is as follows:

1. Top floor: short wires in groups of 1, 2, 3, 4, ... Label the groups A, B, C, D, ... The last group need not be complete.
2. Basement: identify the groups by continuity tests. Number the wires and short them in groups Z, Y, X, W, ...
3. Top floor: remove the shorts. Continuity tests will now uniquely identify all wires. Wire 1 is of course A. Wire 3 is the only wire in group B that has continuity with 1. Its mate will be 2. In group C, only wire 6 connects with 1. Only 5 connects with 2. The remaining wire in C will be 4. And so on for the other groups.

The chart can be extended to the right as far as desired. To determine the procedure for *n* wires, simply cover the chart beyond the *n*th wire.

Figure 64. (Artist: James Egleson.)

5. A continuous line that enters and leaves one of the rectangular spaces must of necessity cross two line segments. Since the spaces labeled A, B, and C in Figure 64 are each surrounded by an odd number of segments, it follows that an end of a line must be inside each if all segments of the network are crossed. But a continuous line has only two ends, so the puzzle is insoluble on a plane surface. This same reasoning applies if the network is on a sphere or on the side of a torus (drawing at lower left). However, the network can be drawn on the torus (drawing at lower right) so that the hole of the torus is *inside* one of the three spaces, A, B, and C. When this is done, the puzzle is easily solved.

   Another approach to this problem was taken by Ken Knowlton and Ronald Graham when both were at Bell Labs. Donald Knuth discussed some small improvements on their ideas in Chapter 30 of his book *Selected Papers on Discrete Mathematics*.

6. Twelve matches can be used to form a right triangle with sides of three, four and five units, as shown in Figure 65. This triangle will have an area of six square units. By altering the position of three matches as shown at right in the illustration, we remove two square units, leaving a polygon with an area of four square units.

Figure 65. (Artist: James Egleson.)

This solution is found in many puzzle books. There are hundreds of other solutions. Elton M. Palmer, Oakmont, Pennsylvania, correlated this problem with the polyominoes of Chapter 13, pointing out that each of the five tetrominoes (figures made with four squares) can provide the base for a large number of solutions. We simply add and subtract the same amount in triangular area to accommodate all 12 matches. Figure 66 depicts some representative samples, each row based on a different tetromino.

Eugene J. Putzer, staff scientist with the General Dynamics Corporation; Charles Shapiro, Oswego, New York; and Hugh J. Metz, Oak Ridge, Tennessee, suggested the star solution shown in Figure 67. By adjusting the width of the star's points you can produce any desired area between 0 and 11.196, the area

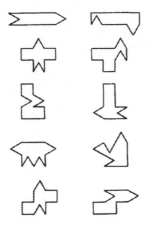

Figure 66. (Artist: James Egleson.)

Figure 67. (Artist: James Egleson.)

of a regular dodecagon, the largest area possible with the 12 matches.

7. Without resorting to calculus, the problem can be solved as follows. Let R be the radius of the sphere. As Figure 68 indicates, the radius of the cylindrical hole will then be the square root of $R^2 - 9$, and the altitude of the spherical caps at each end of the cylinder will be $R - 3$. To determine the residue after the cylinder and caps have been removed, we add the volume of the cylinder, $6\pi(R^2 - 9)$, to twice the volume of the spherical cap, and subtract the total from the volume of the sphere, $4\pi R^3/3$. The volume of the cap is obtained by the following formula, in which A stands for its altitude and r for its radius: $\pi A(3r^2 + A^2)/6$.

When this computation is made, all terms obligingly cancel out except 36π – the volume of the residue in cubic inches. In other words, the residue is constant regardless of the diameter of the hole or the size of the sphere!

The earliest reference I have found for this beautiful problem is on page 86 of Samuel I. Jones's *Mathematical Nuts*, 1932. A two-dimensional analog of the problem appears on page 93 of

Figure 68. (Artist: James Egleson.)

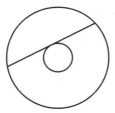

Figure 69

the same volume. Given the longest possible straight line that can be drawn on a circular track of any dimensions (see Figure 69), the area of the track will equal the area of a circle having the straight line as a diameter.

John W. Campbell, Jr., editor of *Astounding Science Fiction*, was one of several readers who solved the sphere problem quickly by reasoning adroitly as follows: The problem would not be given unless it has a unique solution. If it has a unique solution, the volume must be a constant that would hold even when the hole is reduced to zero radius. Therefore the residue must equal the volume of a sphere with a diameter of six inches, namely 36π.

8. At any given instant the four bugs form the corners of a square which shrinks and rotates as the bugs move closer together. The path of each pursuer will therefore at all times be perpendicular to the path of the pursued. This tells us that as A, for example, approaches B, there is no component in B's motion that carries B toward or away from A. Consequently, A will capture B in the same time that it would take if B had remained stationary. The length of each spiral path will be the same as the side of the square: 10 inches.

If three bugs start from the corners of an equilateral triangle, each bug's motion will have a component of 1/2 (the cosine of a 60-degree angle is 1/2) its velocity that will carry it toward its pursuer. Two bugs will therefore have a mutual approach speed of 3/2 velocity. The bugs meet at the center of the triangle after a time interval equal to twice the side of the triangle divided by three times the velocity, each tracing a path that is 2/3 the length of the triangle's side.

The four-bug problem, and its generalizations, are discussed more fully in Chapter 24 of my Book 5.

9. When Jones began to work on the professor's problem he knew that each of the four families had a different number of children and that the total number was less than 18. He further knew that the product of the four numbers gave the professor's house number. Therefore his obvious first step was to factor the house number into four different numbers that together would total less than 18. If there had been only one way to do this, he would have immediately solved the problem. Since he could not solve it without further information, we conclude that there must have been more than one way of factoring the house number.

Our next step is to write down all possible combinations of four different numbers that total less than 18, and obtain the products of each group. We find that there are many cases where more than one combination gives the same product. How do we decide which product is the house number?

The clue lies in the fact that Jones asked if there was more than one child in the smallest family. This question is meaningful only if the house number is 120, which can be factored as $1 \times 3 \times 5 \times 8$, $1 \times 4 \times 5 \times 6$, or $2 \times 3 \times 4 \times 5$. Had Smith answered "No," the problem would remain unsolved. Since Jones did solve it, we know the answer was "Yes." The families therefore contained 2, 3, 4, and 5 children.

This problem was originated by Lester R. Ford and published in the *American Mathematical Monthly*, March 1948, as Problem E776.

CHAPTER THIRTEEN

Polyominoes

Solomon Golomb surprised
Stuffy people who'd not realised
That by adding a square
To the ones that were there
Dominoes could be generalized.
– Anonymous. From the British periodical *Mathematical Pie*,
No. 169, Autumn 2006.

THE TERM "polyomino" was introduced by Solomon W. Golomb,
senior research mathematician in the Jet Propulsion Laboratory
of the California Institute of Technology. In his article "Checker
Boards and Polyominoes" (published in the *American Mathematical
Monthly* in 1954 when Golomb was a 22-year-old graduate student
at Harvard) he defined a polyomino as a "simply connected" set of
squares. By this is meant a set of squares joined along their edges. A
chess player might say, Golomb adds, that they are "rookwise connected," because a rook could travel from any square to any other
square in a finite number of moves. Figure 70 shows a monomino
and all varieties of polyominoes with two, three, and four connected
squares.

There is only one type of domino, two trominoes, and five tetrominoes. When we turn to the pentominoes (five squares) the number jumps to 12. These are shown in Figure 71. Asymmetrical pieces,
which have a different shape when "turned over," are considered as
single types. In all the polyomino recreations to be taken up in this
chapter, asymmetrical pieces may be placed in either of their two
mirror-image forms.

Figure 70. The monomino, the domino, the two trominoes, and the five tetrominoes.

The number of distinct polyominoes of any order is clearly a function of the number of squares in each, but so far no one has succeeded in finding a formula relating the number of *n*-ominoes to *n*. To compute the number of polyominoes of higher orders, one must

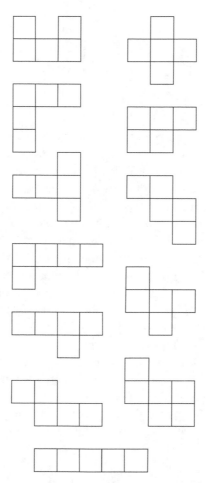

Figure 71. The twelve pentominoes.

fall back on time-consuming procedures when n is large. There are 35 distinct varieties of hexominoes and 108 varieties of heptominoes. This latter figure includes the debatable heptomino shown in Figure 72. In most polyomino recreations, it is best to exclude forms of this type (there are six of them among the octominoes), which have interior "holes." In 2007, the exact number of n-ominoes was known through $n = 28$.

In Chapter 3 (problem 3) we considered a polyomino problem dealing with the placing of dominoes on a mutilated checkerboard. Golomb's article discusses a variety of intriguing similar problems

Figure 72.

involving higher-order polyominoes. It obviously is not possible to cover an 8 × 8 checkerboard with trominoes (because 64 squares are not evenly divisible by 3), but can it be covered with 21 straight trominoes and one monomino? By a clever system of coloring the squares with three colors, Golomb shows this to be possible only when the monomino is placed on one of the four darkened squares in Figure 73. On the other hand, an ingenious induction argument demonstrates that 21 right trominoes and one monomino will cover the 8 × 8 board regardless of where the monomino is placed. It also is possible to cover the board with 16 tetrominoes provided they are all of the same species, the only exception being the skew tetromino, which will not even cover a single edge of the checkerboard. A striped coloring of the board serves to prove that it cannot be covered with 15 L-tetrominoes and one square tetromino; a sawtooth coloring proves it cannot be covered with a square tetromino plus any combination of straight and skew tetrominoes.

Turning to the pentominoes of Figure 70, the question immediately suggests itself: Will these 12 forms, together with one square

Figure 73.

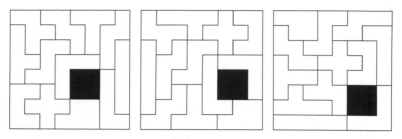

Figure 74. T. R. Dawson's proof.

tetromino, form an 8 × 8 checkerboard? The first published solution of this problem appears in Henry Dudeney's *The Canterbury Puzzles*, 1907. In Dudeney's solution, the square occupies a side position. About 20 years ago, the readers of an obscure British publication called *The Fairy Chess Review* (fairy chess is chess played with unusual rules, boards, or pieces) began experimenting with Dudeney's problem as well as with other pentomino and hexomino patterns. The most interesting results were summarized in the December 1954 issue of the magazine. Much of what follows is drawn from this issue and an unpublished article by Golomb in which he deals with parallel but independently discovered theorems.

T. R. Dawson, founder of *The Fairy Chess Review*, was the first to devise a delightfully simple way to prove that Dudeney's problem can be solved with the square at any position on the board. His three-part solution is depicted in Figure 74. The square tetromino is combined with the L-shaped pentomino to form a 3 × 3 square. By rotating the larger square, the square tetromino can be brought to four different positions in each of the three configurations. Since the entire checkerboard can be both rotated and reflected, it is easy to see that the square tetromino can be placed at any desired spot on the board.

No one knows how many different solutions of this problem there are altogether, but a conservative guess is that there are more than 10,000. In 1958, Dana S. Scott (then a mathematics graduate student at Princeton University, working under contract with the Information Systems Branch of the Office of Naval Research) instructed MANIAC, a digital computer, to search for all possible solutions that

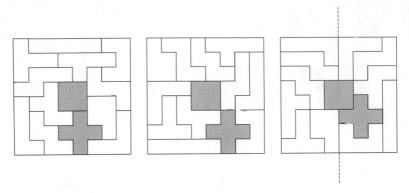

Figure 75.

had the square piece exactly in the center. In an operating time of about three and one-half hours the machine produced an exhaustive list of 65 distinct solutions, not counting additional solutions that can be obtained by rotations and reflections.

In programming the computer, it was convenient to break down the solutions into three categories, each defined by the position of the cross relative to the central square. A solution in each category is shown in Figure 75. The machine found 20 solutions of the first type, 19 of the second type, and 26 of the third.

An inspection of the 65 solutions discloses a number of interesting facts. No solution is possible in which the straight pentomino does not have a long side flush with an edge of the board. (This does not hold for solutions with the square in other positions than the center.) Seven solutions (all in the first and third categories) are without *crossroads*, that is, points where the corners of four pieces meet. The first solution in Figure 75 is of this type. From an artistic standpoint, some polyomino experts have considered crossroads to be blemishes in a design. The third solution of Figure 75 illustrates another interesting feature: a straight line on which the pattern can be folded in half. There are 11 patterns of this type, all in the third category and none free of crossroads.

If the square tetromino is discarded and four disconnected unit squares are left open, the 8 × 8 checkerboard can still be formed in a large number of artistic ways. Three such patterns are shown in Figure 76. It also is possible to fit the 12 pentominoes into rectangles

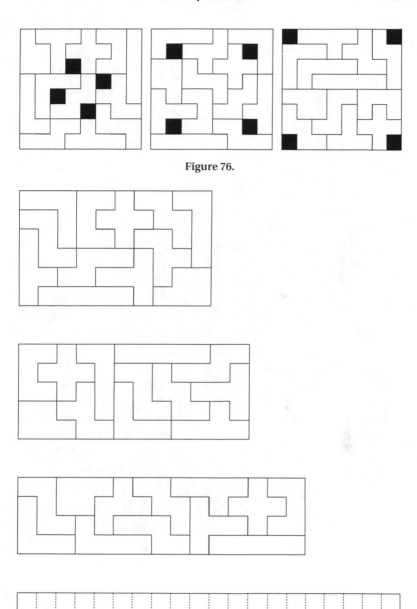

Figure 76.

Figure 77. Pentomino rectangles.

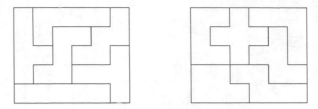

Figure 78. Triplication patterns.

that are 6 × 10, 5 × 12, 4 × 15, and 3 × 20 (see Figure 77). The 3 × 20 rectangle, by all odds the most difficult, is left for the interested reader to construct. It has only two distinct solutions, not counting rotations and reflections.

Note that the 5 × 12 rectangle in Figure 76 is shown here with a solution that contains a 5 × 7 and a 5 × 5 rectangle. Several readers discovered the two 5 × 6 rectangles shown in Figure 78, which can be put together to make either a 5 × 12 or a 6 × 10 rectangle.

Raphael M. Robinson, professor of mathematics at the University of California, recently proposed what he calls "the triplication problem." You select one pentomino and then use nine of the remaining ones to form a large scale-model of the chosen piece. The model will be three times higher and wider than the small one. Joseph B. Tucker, rector of Trinity Episcopal Church in Clarksville, Tennessee, independently hit on the triplication problem after reading this department's discussion of pentominoes. He sent in many excellent solutions, including the two shown in Figure 79. The triplication problem can be solved for each of the 12 pieces.

Figure 79.

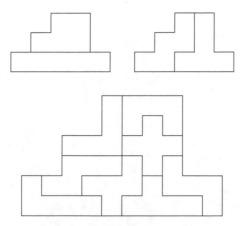

Figure 80. "Double double" pattern.

Somewhat similar problems were proposed by other readers. Harry Brueggemann of San Marino, California, suggested what he termed the "double double problem." You first form any desired shape with two pentominoes. You duplicate it with two other pieces. Finally, the remaining eight pieces are used to form the same shape but twice as large. Figure 80 shows a typical solution. Paul J. Slate of West Orange, New Jersey, proposed using all 12 pieces to make a 5 × 13 rectangle with a hole in the shape of one of the pieces. It can be solved with a hole in the form of each pentomino. One such solution is depicted in Figure 81.

Another interesting pentomino problem, proposed in *The Fairy Chess Review* by H. D. Benjamin, is shown in Figure 82. The 12 pentominoes will exactly cover a cube that is the square root of 10 units

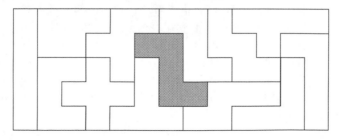

Figure 81. A pentomino cube.

Figure 82. The pentomino game.

on the side. The cube is formed by folding the pattern along the dotted lines.

What is the minimum number of pentominoes that can be placed on a checkerboard in such a way that it is impossible to place any of the remaining pentominoes on the board? This intriguing question is asked by Golomb, and he says the answer is five. Figure 83 shows one such configuration. This problem suggested to Golomb a fascinating competitive game that can be played on a checkerboard with large cardboard pentominoes cut to fit accurately over the board's squares. (The reader is invited to make such a set, not only to enjoy the game, but also to solve pentomino problems and create new ones.)

Two or more players take turns in choosing a single pentomino and placing it wherever they wish on the board. The pieces have

Figure 83.

no "top" or "bottom" faces. As in all problems mentioned in this article, asymmetrical pieces may be used with either side up. The first player who is unable to place a piece is the loser.

Golomb writes:

> The game will last at least five and at most 12 moves, can never result in a draw, has more possible openings than chess, and will intrigue players of all ages. It is difficult to advise what strategy should be followed, but there are two valuable principles:
>
> 1. Try to move in such a way that there will be room for an even number of pieces. (This assumes only two are playing.)
> 2. If you cannot analyze the situation, do something to complicate the position, so that the next player will have even more difficulty analyzing it than you did.

Since the 35 hexominoes have a total area of 210 squares, one thinks immediately of arranging them to form a rectangle that could be 3 × 70, 5 × 42, 6 × 35, 7 × 30, 10 × 21, or 14 × 15. I seriously considered offering $1,000 to the first reader who succeeded in constructing one of these six rectangles, but the appalling thought of hours that might be wasted on the challenge forced me to relent. All such efforts are doomed to failure. Golomb's proof of this is a striking example of the use of two powerful tools of combinatorial geometry. This is a little known branch of mathematics, though it has many practical applications to engineering design problems involving standard components that must be fitted together in the most efficient manner. The tools are (1) the use of contrasting colors to aid one's mathematical intuition and (2) the principle of "parity check" based on the combinatorial properties of odd and even numbers.

We begin the proof by coloring our desired rectangles with alternating black and white squares like a checkerboard. In each case, the rectangle clearly must contain 105 black squares and 105 white – an odd number for each.

Turning our attention to the 35 hexominoes, we discover that 24 of them will always cover three black squares and three white – an odd number for each. There is an even number of these "odd hexominoes," and since even times odd is even, we know that all 24 of them will cover an even number of squares of each color.

The remaining 11 hexominoes are of such a shape that each must cover four squares of one color and two of the other – an even number for each. There is an odd number of these "even hexominoes," but again, since even times odd is even, we know that these 11 pieces also will cover an even number of squares of each color. (Figures 84 and 85 divide the 35 hexominoes into even and odd groups.) Finally, since even plus even is even, we conclude that the 35 hexominoes together will cover an even number of black squares and an even number of white squares. Unfortunately each rectangle contains 105 squares of each color. This is an odd number. No rectangle, therefore, can be covered by the 35 hexominoes.

"There is a lesson in plausible reasoning to be learned from these problems," Golomb concludes.

> Given certain basic data, we labor long and hard to fit them into a pattern. Having succeeded, we believe the pattern to be the only one that "fits the facts"; indeed, that the data are merely manifestations of the beautiful, comprehensive whole. Such reasoning has been used repeatedly in religion, in politics, even in science. The pentominoes illustrate that many different patterns may be possible from the same "data," all equally valid, and the nature of the pattern we end up with is determined more by the shape we are looking for than by the data at hand. It is also possible that for certain data [as in the hexomino problem explained previously], no pattern of the type we are conditioned to seek may exist.

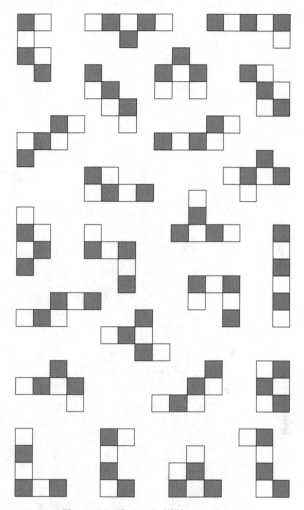

Figure 84. The 24 "odd" hexominoes.

ADDENDUM

For readers who may wish to experiment with hexomino patterns, I add here (Figures 86 and 87) two striking designs reproduced from *The Fairy Chess Review.* Each is formed with the complete set of 35 hexominoes. Patterns using the entire set cannot be made unless

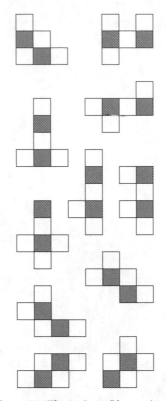

Figure 85. The 11 "even" hexominoes.

Figure 86. A hexomino pattern.

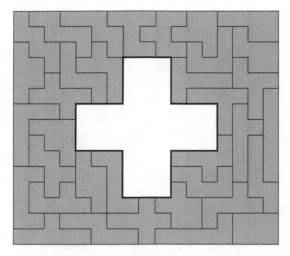

Figure 87. Another hexomino design.

a checkerboard coloring of the squares shows an excess of squares of one color in the amount of 2, 6, 10, 14, 18, or 22.

A box of colored plastic pentominoes was marketed in 1957 by Tryne Products, Inc. under the trade name of Hexed.

AFTERWORD, 1988

So many articles have appeared about polyominoes, polycubes (joined cubes), polyhexes (joined hexagons), and tetrominoes (joined right-angle isosceles triangles) that I have made no attempt to cover this literature in my bibliography. Readers are referred to the references cited in chapters about polyominoes in my later book collections of columns, and to issues of the *Journal of Recreational Mathematics* in which dozens of articles and problems involving polyominoes have been published. Sets of polyominoes and their cousins appear on the market from time to time in various countries. A handsome wooden set of solid pentominoes (the 12 pentominoes with unit thickness) is currently available from Kadon Enterprises, 1227 Lorene Drive, Pasadena, MD 21122. They can be used both for problems on the plane and in three dimensions.

POSTSCRIPT

Since I introduced the mathematical community to Sol Golomb's polyominoes, the literature involving these fascinating shapes has become enormous. Golomb's 1965 book *Polyominoes* was reissued in a much revised edition by Princeton University Press in 1994. It has a bibliography of earlier references. George Edward Martin's *Polyominoes: A Guide to Puzzles and Problems in Tiling*, was published by the Mathematical Association of America in 1991.

Dozens of problems involving polyominoes are scattered throughout chapters in later books of my *Scientific American* columns, and even a larger number of such problems have appeared in issues of the *Journal of Recreational Mathematics*. Polyominoes' cousins, the polyiamonds, polyhexes, and polyaboloes, are topics of chapters in later books in this series, as are the solids called polycubes of which Piet Hein's Soma Cube is the most famous such puzzle.

Naoaki Takashima, of Tokyo, at an Atlanta Gathering for Gardner (March 2004), lectured on the task of covering a unit cube with the 12 pentominoes and distributed a privately published monograph titled "Pentomino Cube Solutions." It covers recent computer programs that have found tens of thousands of different solutions, the exact number depending on how "different" is defined. C. J. Bouwkamp, of Eindhoven University of Technology, in the Netherlands, found 26,358,584 solutions of which 284,402 were called "nice." A nice solution is one in which none of the three cells at a cube's corners are occupied by the same pentomino. His results are given in his university's 1998 Report 97-WSK-01. They were verified by Takashima's program, implemented by Kayoko Sakai.

In the December 1990 issue of the Dutch periodical *Cubism for Fun*, I posed the task of covering a unit cube with n congruent polygons, n greater than 1. The case of $n = 2$ is easily solved by two 3×1 rectangles. Because each rectangle can be divided into n identical rectangles, it follows that the cube can be tiled with any *even* number of congruent rectangles.

The $n = 3$ case is also easily solved by three 2×1 rectangles. Again, each rectangle can be cut into n congruent rectangles; therefore, the cube can be covered by n congruent rectangles when n is a multiple of 3.

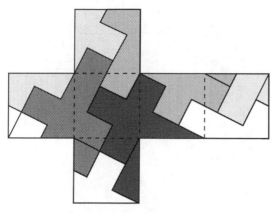

Figure 88. Tiling the cube with five cross hexominoes. (Artist: Harold Jacobs.)

The case of $n = 5$ can be solved by the Latin cross hexomino as shown in Figure 88. This pattern appeared in *Mathematics Magazine*, Vol. 50 (1977), pp. 158–178. In 1998, Bouwkamp reported that his computer program found 1,054 solutions with twelve identical pentominoes of which 164 were "nice." Pieter Torbijn discusses the general problem of tiling a cube with identical polyominoes in *Cubism for Fun*, Vol. 58, July 2002; Vol. 59, November 2002; and Vol. 61, July 2003. The latter issue also has an article on the general topic by Markus Götz.

In my *Cubism for Fun* article I offered $50 to anyone who could solve the next higher case, $n = 7$. To my great surprise, the prize was won by Anneke Treep who found four different strips, each of which can be sliced into n congruent parts! This, of course, solves the general problem for all n!

Figure 89 shows how the first strip can be divided into n identical polygons. The other three strips can be divided in a similar way. Treep explained this in her article "Covering a Cube" in *Cubism for Fun*, Vol. 27, December 1991, pp. 16–17. My piece in which I offered the prize, followed by pictures showing Treep's solution for any n, is reprinted in *Gardner's Workout* (A K Peters, 2001), Chapter 8.

The Spring 2007 issue of the British periodical *Pie* describes two polyomino cousins, the polyiamonds (made with equilateral triangles) and the polyaboloes (made with right triangles). Both families are topics of chapters in later books of this series.

Figure 89. Covering a cube with *n* congruent polygons. (Artist: Harold Jacobs.)

Suppose, *Pie* continues, we consider a new family called *poly-cornoes*. They are made with unit squares, joined at their *corners* instead of their sides as shown in Figure 91. *Pie* asked: How many different shapes (not counting rotations and reflections) can be made by joining *n* squares?

The surprising answer, proved by the pieces shown in Figure 90, is that the number of polycornoes of *n* squares is exactly the same

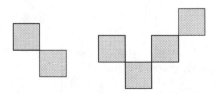

Figure 90. (Artist: Harold Jacobs.)

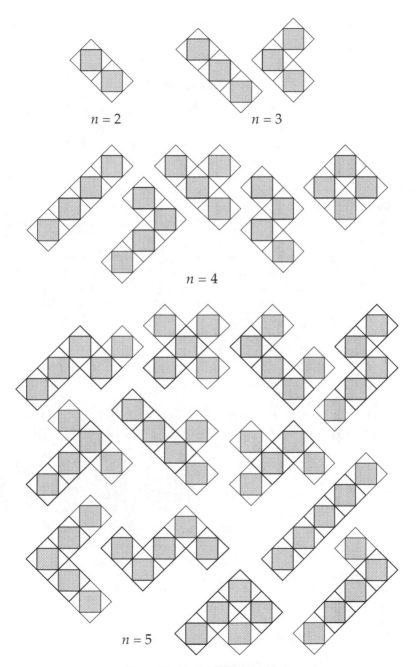

$n = 2$

$n = 3$

$n = 4$

$n = 5$

Figure 91. (Artist: Harold Jacobs.)

as the number of polyominoes of n squares! In other words, the two families are duals.

BIBLIOGRAPHY

"Checkerboards and Polyominoes." Solomon W. Golomb in *American Mathematical Monthly* 61 (December 1954): 675–682.

"Dissection." Thirty-eight pentomino and hexomino constructions. W. Stead in *Fairy Chess Review* 9 (December 1954): 2–4.

"Programming a Combinatorial Puzzle." Dana S. Scott. Technical Report No. 1, 10 June 1958, Department of Electrical Engineering, Princeton University.

Polyominoes. Solomon W. Golomb. Scribner's, 1965. Rev. ed. Princeton University Press, 1994.

"Polyominoes and Fault-Free Rectangles." Martin Gardner in *New Mathematical Diversions from Scientific American.* Chapter 13. Simon and Schuster, 1966. University of Chicago Press reprint, 1983.

"Tiling with Polyominoes." Solomon W. Golomb in *Journal of Combinatorial Theory* 1 (September 1966): 280–286.

"Tiling with Sets of Polyominoes." Solomon W. Golomb in *Journal of Combinatorial Theory* 9 (July 1970): 60–71.

"Tesselations of Polyominoes." A. W. Bell in *Mathematical Reflections*, edited by members of the Association of Teachers of Mathematics. Cambridge, 1970.

Imperial Earth. Arthur C. Clarke. Harcourt, 1976. A set of pentominoes is basic to the plot, serving as a symbol of life's combinatorial possibilities.

"Polyominoes and Rectification." Martin Gardner in *Mathematical Magic Show.* Chapter 13. Knopf, 1977.

"Help, I'm a Pentomino Addict!" Arthur C. Clarke in *Ascent to Orbit.* Wiley, 1984.

"Tiling with Polyominoes, Polyiamonds, and Polyhexes." Martin Gardner in *Time Travel and Other Mathematical Bewilderments.* Chapter 14. W. H. Freeman, 1988.

Polyominoes. George Edward Martin. Mathematical Association of America, 1991.

Pentominoes. John Millington. Tarquin, 1993.

"My Life Among the Polyominoes." David Klarner in *Mathematical Recreations*, ed. David Klarner. Dover, 1998.

CHAPTER FOURTEEN

Fallacies

A MATHEMATICAL PARADOX CAN BE DEFINED as a mathematical truth so startling that it is difficult to believe even after every step of its proof has been verified. Mathematical fallacies are equally astonishing assertions, but unlike mathematical paradoxes their proofs contain subtle errors. Every branch of mathematics, from simple arithmetic to topological set theory, has its share of these counterfeit arguments. The better ones are of course those with the most incredible conclusions and the best-camouflaged errors. Euclid devoted an entire book to geometrical fallacies, but his manuscript did not survive, so we can only speculate on what this lost classic of recreational mathematics may have contained.

The following seven fallacies have been selected for their variety and interest. They will not be explained, but the reader may find it pleasant and instructive to seek out their errors.

Our first fallacy is an exceedingly elementary one. We shall introduce it by way of an amusing paradox that David Hilbert, the great German mathematician, liked to employ to illustrate one of the peculiar properties of aleph-null, the smallest of the transfinite numbers. It seems that the manager of a celestial hotel with an infinite number of rooms, all occupied, wishes to accommodate a new guest. He does so by moving each occupant to a room with the next highest number, thereby vacating Room 1. What can he do if an infinite number of new guests arrive? The undismayed manager simply shifts each occupant to a room that has a number twice as large as that of his first room; the guest in Room 1 goes to Room 2, the guest in 2 goes to 4, 3 to 6, 4 to 8, and so on. This opens up all the odd-numbered rooms, which will accommodate every one.

But is it really necessary that the number of occupied rooms be infinite before additional guests can be accommodated? The following doggerel from a late nineteenth-century British magazine tells how a clever innkeeper with nine empty rooms had no difficulty in providing separate lodgings for each of ten travelers.

> Ten weary, footsore travelers,
> All in a woeful plight,
> Sought shelter at a wayside inn
> One dark and stormy night.
>
> "Nine rooms, no more," the landlord said,
> "Have I to offer you.
> To each of eight a single bed,
> But the ninth must serve for two."
>
> A din arose. The troubled host
> Could only scratch his head,
> For of those tired men no two
> Would occupy one bed.
>
> The puzzled host was soon at ease –
> He was a clever man –
> And so to please his guests devised
> This most ingenious plan.
>
> In room marked A two men were placed,
> The third was lodged in B,
> The fourth to C was then assigned,
> The fifth retired to D.
>
> In E the sixth he tucked away,
> In F the seventh man,
> The eighth and ninth in G and H,
> And then to A he ran,
>
> Wherein the host, as I have said,
> Had laid two travelers by;
> Then taking one – the tenth and last –
> He lodged him safe in I.
>
> Nine single rooms – a room for each –
> Were made to serve for ten;
> And this it is that puzzles me
> And many wiser men.

A slightly more sophisticated fallacy is the following algebraic proof that any number a is equal to a smaller number b.

$$a = b + c$$

Multiply both sides by $a - b$ to obtain

$$a^2 - ab = ab + ac - b^2 - bc$$

Subtract ac from each side:

$$a^2 - ab - ac = ab - b^2 - bc$$

Factor:

$$a(a - b - c) = b(a - b - c)$$

Divide each side by $a - b - c$ to get

$$a = b$$

Manipulation of the imaginary number i (the square root of -1) has many pitfalls, as witnessed by the following tantalizing proof:

$$\sqrt{-1} = \sqrt{-1}$$

$$\sqrt{\frac{1}{-1}} = \sqrt{\frac{-1}{1}}$$

$$\frac{\sqrt{1}}{\sqrt{-1}} = \frac{\sqrt{-1}}{\sqrt{1}}$$

$$\sqrt{1} \times \sqrt{1} = \sqrt{-1} \times \sqrt{-1}$$

$$1 = -1$$

In plane geometry most fallacies hinge on an improperly constructed diagram. Consider for example this perplexing demonstration that the front side of a polygon cut out of a piece of paper has an area that differs from that of the back side. The demonstration was devised by L. Vosburgh Lyons, a New York neuropsychiatrist, to exploit a curious principle recently discovered by Paul Curry, also of New York.

First draw on a sheet of graph paper the 60-square-unit triangle shown in Figure 92. Cut along the lines to make six pieces and then color the back of each piece. If all six pieces are turned over and a

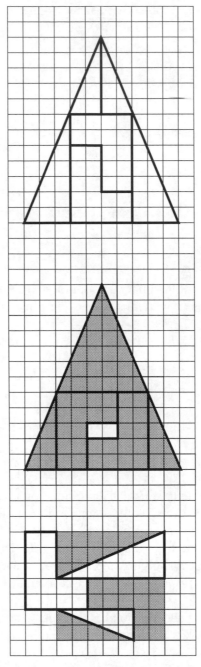

Figure 92. The Curry triangle. (Artist: Bunji Tagawa.)

colored triangle formed as shown in the middle of the illustration, it will be found that the triangle has developed a hole of two square units. In other words, its area has shrunk to 58 square units. If we turn three pieces so that their white sides are uppermost, leaving three colored pieces, we can form the figure shown at the bottom of the illustration. This has the in-between area of 59 square units. Something is obviously wrong here, but what?

Probability theory swarms with plausible but specious lines of reasoning. Suppose you have just met your friend Jones and each of you is wearing a necktie that you received as a Christmas present. You begin to argue over which of you received the more expensive tie. You and Jones finally agree to settle the matter by visiting the store where both ties were bought and checking their value. The man who wins (i.e., has the most expensive tie) must give his tie to the loser as a consolation.

This is how you reason: "The chances that I will win the argument or lose it are equal. If I win, I will be poorer by the value of this tie I am wearing. But if I lose, I am sure to gain a more expensive tie. Therefore, the contest is clearly to my advantage."

Of course Jones can reason in exactly the same way. How can a bet be favorable to both parties?

One of the most surprising paradoxes of topology is the fact that a torus (a doughnut-shaped surface) can be turned inside out through a hole in its side by stretching the surface without tearing it. There is no question about this. When the steps in the process were depicted in *Scientific American* for January 1950, a New Jersey engineer actually shipped the magazine in an inner tube that he had reversed. But if this can be done, then an even more remarkable fact seems to emerge.

On the outside of a torus paint the ring at right in the upper illustration of Figure 93. On the inside of the same torus paint a second ring. These two closed curves are clearly linked. The torus is now turned inside out through the hole. As the bottom illustration shows, this moves the first ring to the inside and the second ring to the outside. The rings are no longer linked! This obviously violates a fundamental topological law that states that two linked curves cannot be separated without breaking one curve and passing the other through the break.

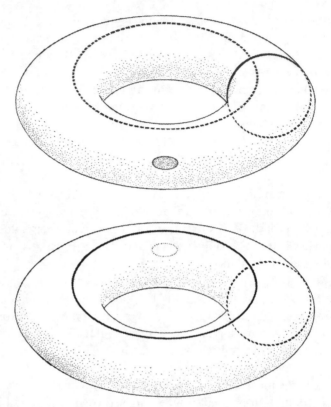

Figure 93. Two linked rings appear to unlink when torus is turned inside out through hole in its side. (Artist: Bunji Tagawa.)

Our final fallacy, which draws on elementary number theory, concerns "interesting" versus "uninteresting" numbers. Numbers can of course be interesting in a variety of ways. The number 30 was interesting to George Moore when he wrote his famous tribute to "the woman of 30," the age at which he believed a married woman was most fascinating. To a number theorist, 30 is more likely to be exciting because it is the largest integer such that all smaller integers with which it has no common divisor are prime numbers. The number 15,873 is intriguing because if you multiply it by any digit and then by 7, the result will consist entirely of repetitions of the chosen digit. The number 142,857 is even more fascinating. Multiply it by any digit from 1 through 6 and you get the same six digits in the same cyclic order.

The question arises: Are there any uninteresting numbers? We can prove that there are none by the following simple steps. If there are dull numbers, we can then divide all numbers into two sets – interesting and dull. In the set of dull numbers, there will be only one number that is the smallest. Since it is the smallest uninteresting number it becomes, *ipso facto*, an interesting number. We must therefore remove it from the dull set and place it in the other. But now there will be another smallest uninteresting number. Repeating this process will make any dull number interesting.

ADDENDUM

Two readers favored me with ninth stanzas for the poem about the ten weary, footsore travelers. (This poem appeared, by the way, in the magazine *Current Literature*, Vol. 2, April 1889, page 349. No author's name is given, but it is credited to the *Pittsburgh Bulletin*, no date. The paradox is much older than the poem; still it would be interesting to know who gave it this poetic form.) Ralph W. Allen of Los Angeles wrote:

> I had not heard the din that night
> As number ten raised hue and cry –
> 'Twas number two – not number ten –
> That bedded down in room marked I.

John F. Mooney, of the Ebasco International Corporation, New York, New York, exposed the fallacy this way:

> If we reflect on what he's done
> We'll see we're not insane.
> Two men in A, he's counted one,
> Not once, but once again.

The fallacy that disturbed most readers was the one about the inside-out torus. It is true that the torus can be reversed, but the reversal changes the "grain," so to speak, of the torus. As a result, the two rings exchange places and remain linked. Several readers made excellent models by cutting off the upper part of a sock, then sewing the ends of the upper part together to make the torus. The rings consisted of thread, in two contrasting colors, stitched to the outside and inside of the cloth torus. Such a torus reverses easily through a hole in the side, demonstrating most effectively exactly what happens to the rings.

The necktie paradox is fully discussed in Maurice Kraitchnik's *Mathematical Recreations*, a Dover book.

The closing "proof" that no numbers are uninteresting prompted the following telegram from Dave Engle, at the College of Puget Sound, Tacoma, Washington:

PER JANUARY SCIENTIFIC AMERICAN SUGGEST THAT JUST SHORT OF INFINITY YOU CEASE SNIPPING OFF AND REMOVING DULL NUMBERS. AT LEAST SAVE ONE FOR INTEREST'S SAKE!

POSTSCRIPT

Variations on the Curry triangle paradox, and related "geometrical vanishes," can be found in my *Mathematics, Magic, and Mystery* (Dover, 1956). My major contribution to such flummery, a square with a large hole that appears in the center when pieces are reassembled a different way, is in Chapter 10 of my Book 3.

Many variations on Curry rectangles have been marketed as magic tricks and advertising premiums. Magic shops sell a large playing card that develops a hole when the pieces are turned over and put together again, and a dollar bill that loses the face of George Washington when its four pieces are reassembled. There was even a pornographic version marketed many years ago in Japan that featured a picture of a nude young woman on one side.

Geometric fallacies, such as a classic proof that all triangles are isosceles, and that a right angle equals an obtuse angle, are covered in Chapter 6 of Book 10.

BIBLIOGRAPHY

Riddles in Mathematics: A Book of Paradoxes. Eugene P. Northrop. Van Nostrand, 1944.
Paradoxes and Common Sense. Aubrey Kempner. Van Nostrand, 1959.
Fallacies in Mathematics. E. A. Maxwell. Cambridge University Press, 1959.

Lapses in Mathematical Reasoning. V. M. Bradis, V. L. Minkovskii, and A. K. Kharcheva. Pergamon, 1963.

Mistakes in Geometric Proofs. Ya. S. Dubnov. Heath, 1963.

"The Disappearing Man and Other Vanishing Paradoxes." Mel Stover in *Games* (November–December 1980): 14–18.

"Geometrical Fallacies." Martin Gardner in *Wheels, Life, and Other Mathematical Amusements.* W. H. Freeman, 1983.

THE NECKTIE PARADOX

"The Other Person's Envelope Is Always Greener." B. Nalebuff in *Journal of Economic Perspectives* 3 (1989): 171–181.

"The Box Problem: To Switch or Not to Switch." S. J. Brams and D. M. Kilgour in *Mathematics Magazine* 68 (1995): 27–34.

"Elusive Optimality in the Box Problem." N. M. Blachman and D. M. Kilgour in *Mathematics Magazine* 74 (2001): 171–181.

"Risk Management and the Two-Envelope Paradox." Edwin F. Meyer in *Journal of Recreational Mathematics* 31 (2002–2003): 275–279.

"One Observation Behind Two-Envelope Puzzles." Dov Samet, Iddo Samet, and David Schmeidler in *American Mathematical Monthly* 111 (April 2004): 347–351.

"On the Two-Box Paradox." Robert A. Agnew in *Mathematics Magazine* 77 (October 2004): 302–308.

CHAPTER FIFTEEN

Nim and Tac Tix

ONE OF THE OLDEST AND MOST ENGAGING OF all two-person mathematical games is known today as Nim. Possibly Chinese in origin, it is sometimes played by children with bits of paper, and by adults with pennies on the counter of a bar. In the most popular version of the game 12 pennies are arranged in three horizontal rows as shown in Figure 94.

The rules are simple. The players alternate in removing one or more coins provided they all come from the same horizontal row. Whoever takes the last penny wins. The game can also be played in reverse: Whoever takes the last penny loses. A good gamester soon discovers that in either form of the game he can always win if one of his moves leaves two rows with more than one penny in a row and the same number in each; or if the move leaves one penny in one row, two pennies in a second row and three in a third. The first player has a certain win if on his first move he takes two pennies from the top row and thereafter plays "rationally."

There is nothing startling about the foregoing analysis, but around the turn of the century an astonishing discovery was made about the game. It was found that it could be generalized to any number of rows with any number of counters in each, and that an absurdly simple strategy, using binary numbers, would enable anyone to play a perfect game. A full analysis and proof was first published in 1901 by Charles Leonard Bouton, associate professor of mathematics at Harvard University. It was Bouton, incidentally, who named the game Nim, presumably after the archaic English verb meaning to take away or steal.

In Bouton's terminology every combination of counters in the generalized game is either "safe" or "unsafe." If the position left by

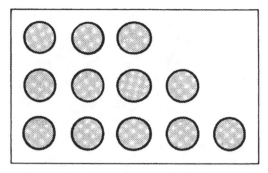

Figure 94. Twelve counters are arranged for a "3, 4, 5" game of Nim.

a player after his move guarantees a win for that player, the position is called safe. Otherwise it is unsafe. Thus in the "3, 4, 5" game previously described the first player leaves a safe position by taking two pennies from the top row. Every unsafe position can be made safe by a proper move. Every safe position is made unsafe by *any* move. To play rationally, therefore, a player must move so that every unsafe position left to him is changed to a safe position.

To determine whether a position is safe or unsafe, the numbers for each row are written in binary notation. If each column adds up to zero or an even number, then the position is safe; otherwise, it is not.

There is nothing mysterious about the binary notation. It is merely a way of writing numbers by sums of the powers of two. The chart of Figure 95 shows the binary equivalents of the numbers 1 through 20. You will note that each column, as you move from right to left, is headed by a successively higher power of two. Thus the binary number 10101 tells us to add 16 to 4 to 1, giving us 21 as its equivalent in the decimal system, based on the powers of 10. To apply the binary analysis to the 3, 4, 5 starting position of Nim, we first record the rows in binary notation as follows:

	4	2	1
3		1	1
4	1	0	0
5	1	0	1
Totals	2	1	2

	16	8	4	2	1
1					1
2				1	0
3				1	1
4			1	0	0
5			1	0	1
6			1	1	0
7			1	1	1
8		1	0	0	0
9		1	0	0	1
10		1	0	1	0
11		1	0	1	1
12		1	1	0	0
13		1	1	0	1
14		1	1	1	0
15		1	1	1	1
16	1	0	0	0	0
17	1	0	0	0	1
18	1	0	0	1	0
19	1	0	0	1	1
20	1	0	1	0	0

Figure 95. Table of binary numbers for playing Nim.

The middle column adds up to 1, an odd number, telling us that the combination is unsafe. It can therefore be made safe by the first player. He does so, as explained, by taking two pennies from the top row. This changes the top binary number to 1, thereby eliminating the odd number from the column totals. The reader will discover by trying other first moves that this is the only one that makes the position safe.

An easy way to analyze any position, provided there are no more than 31 counters in one row, is to use the fingers of your left hand as a binary computer. Suppose the game begins with rows of 7, 13, 24, and 30 counters. You are the first player. Is the position safe or unsafe? Extend all five fingers of your left hand, palm toward you. The thumb registers units in the 16 column; the index finger, those in the 8 column; the middle finger, the 4 column; the ring finger, the 2 column; the little finger, the 1 column. To feed 7 to your computer, first bend down the finger representing the largest power of 2 that will go into 7. It is 4, so you bend your middle finger. Continue adding powers of two, moving to the right across your hand, until the total is 7. This is of course reached by bending the middle, ring and little fingers. The remaining three numbers – 13, 24, and 30 – are fed to your computer in exactly the same way except that any bent finger involved in a number is raised instead of lowered.

Regardless of how many rows there are in the game, if you finish this procedure with all your fingers raised, then the position is safe. This means that your move is sure to make it unsafe, and that you are certain to lose against any player who knows as much about Nim as you do. In this example, however, you finish with first and second fingers bent, telling you that the position is unsafe, and that you can win if you make a proper move. Because there are many more unsafe combinations than safe ones, the odds greatly favor the first player when the starting position is determined at random.

Now that you know that 7, 13, 24, 30 is unsafe, how do you find a move that will make it safe? This is difficult to do on your fingers, so it is best to write down the four binary numbers as follows:

	16	8	4	2	1
7			1	1	1
13		1	1	0	1
24	1	1	0	0	0
30	1	1	1	1	0
Totals	2	3	3	2	2

Note the column farthest to the left that adds up to an odd number. Any row with a unit in this column can be altered to make the position safe. Suppose you wish to remove a counter or counters from the second row. Change the first unit to 0; then adjust the remaining figures on the right so that no column will add up to an odd number. The only way to do this is to change the second binary number to 1. In other words, you remove all counters except one from the second row. The other two winning moves would be to take 4 from the third row or 12 from the last row.

It is helpful to remember that you can always win if you leave two rows with the same number of counters in each. From then on, simply move each time to keep the rows equal. This rule, as well as the preceding binary analysis, is for the normal game in which you win by taking the last counter. Happily only a trivial alteration is required to adopt this strategy to the reverse game. When the reverse game reaches a point (as it must) at which only one row has more than one counter, you must take either all or all but one counter from that row so as to leave an odd number of one-unit rows. Thus if the board shows 1, 1, 1, 3, you take the entire last row. If it shows 1, 1, 1, 1, 8, you take seven from the last row. This modification of strategy occurs only on your final move, when it is easy to see how to win.

Since digital computers operate on the binary system, it is not difficult to program such a computer to play a perfect game of Nim or to build a special machine for this purpose. Edward U. Condon, the former director of the National Bureau of Standards who is now head of the physics department at Washington University in St. Louis, was a co-inventor of the first such machine. Patented in 1940 as the Nimatron, it was built by the Westinghouse Electric Corporation and exhibited in the Westinghouse building at the New York World's Fair. It played 100,000 games and won 90,000. Most of its defeats were administered by attendants demonstrating to skeptical spectators that the machine could be beaten.

In 1941 a vastly improved Nim-playing machine was designed by Raymond M. Redheffer, now assistant professor of mathematics at the University of California at Los Angeles. Redheffer's machine has the same capacity as Condon's (four rows with as many as seven counters in each), but where Nimatron weighed a ton and required costly relays, Redheffer's machine weighs 5 pounds and uses only four rotary switches. Ten years later, a Nim-playing robot called Nimrod was exhibited at the Festival of Britain in 1951 and later at the Berlin Trade Fair. According to an account by A. M. Turing (in Chapter 25 of *Faster Than Thought*, edited by B. V. Bowden, 1953), the machine was so popular in Berlin that visitors "entirely ignored a bar at the far end of the room where free drinks were available, and it was necessary to call out special police to control the crowds. The machine became even more popular after it had defeated the economics minister, Dr. Erhard, in three games."

Among many variations of Nim that have been fully analyzed, one proposed in 1910 by the American mathematician Eliakim H. Moore is of special interest. The rules are the same as they are for regular Nim except that players are permitted to take from any number of rows not exceeding a designated number k. Surprisingly, the same binary analysis holds, provided a safe position is defined as one in which every column of the binary numbers totals a number evenly divisible by $(k + 1)$.

Other variations of Nim seem not to have any simple strategy for rational play. To my mind the most exciting of these as yet unanalyzed versions was invented about 10 years ago by Piet Hein of Copenhagen. (Piet Hein is the inventor of Hex, a topological game discussed in Chapter 8.)

In Piet Hein's version, called Tac Tix in English-speaking countries and Bulo in Denmark, the counters are arranged in square formation as shown in Figure 96. Players alternately take counters, but they may be removed from any horizontal or vertical row. They must always be adjoining counters with no gaps between them. For example, if the first player took the two middle counters in the top row, his opponent could not take the remaining counters in one move.

Tac Tix must be played in reverse form (the player who takes the last counter loses) because of a simple strategy that renders the normal game trivial. On squares with an odd number of counters on each side the first player wins by taking the center counter and then playing symmetrically opposite his opponent. On squares with an even number of counters on each side the second player wins by playing symmetrically from the outset. No comparable strategy is known for playing the reverse game, although it is not difficult to show that on a 3 × 3 board the first player can win by taking the center counter or a corner counter, or all of a central row or column.

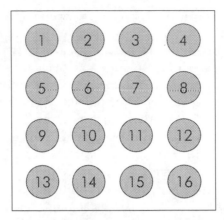

Figure 96. Piet Hein's game of Tac Tix.

The clever principle behind Tac Tix, that of intersecting sets of counters, has been applied by Piet Hein to many other two- and three-dimensional configurations. The game can be played, for example, on triangular and hexagonal boards, or by placing the counters on the vertices and intersections of a pentagram or hexagram. Intersections of closed curves may also be used; here all counters lying on the same curve are regarded as being in the same "row." The square form, however, combines the simplest configuration with maximum strategic complexity. It is difficult enough to analyze even in the elementary 4 × 4 form, and of course as the squares increase in size the game's complexity rapidly accelerates.

A superficial analysis of the game suggests that symmetry play might ensure a win for the second player in a 4 × 4 game, with only a trivial modification on his last move. Unfortunately, there are many situations in which symmetry play will not work. For example, consider the following typical game in which the second player adopts a symmetry strategy.

	First Player	Second Player
1.	5–6	11–12
2.	1	16
3.	4	13
4.	3–7 (wins)	

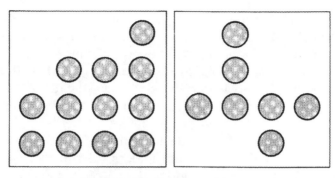

Figure 97. Two problems of Tac Tix.

In this example, the second player's initial move is a fatal one. After his opponent responds as indicated, the second player cannot force a win even if he departs from symmetry on all his succeeding moves.

The game is much more complex than it first appears. In fact, it is not yet known whether the first or second player can force a win even on a 4 × 4 board from which the four corner pieces have been removed. As an introduction to the game, try solving the two Tac Tix problems (devised by Piet Hein) pictured in Figure 97. On each board you are to find a move that ensures a win. Perhaps some industrious reader can answer the more difficult question: Who has a win on the 4 × 4 board, the first or second player?

ADDENDUM

Seville Chapman, director of the physics division of the Cornell Aeronautical Laboratory, Inc., at Cornell University, sent me a wiring diagram for a well-thought-out portable Nim machine that he built in 1957. It weighs 34 ounces, using three multideck rotary switches to handle three rows of 4 to 10 counters each. By taking the first move, the machine can always win. There is a rather pretty way to prove this. If we record the three rows in the matrix form previously described, it is clear that each row must have a 1 in either the 8 or 4 column, but not in both. (The two spaces cannot be empty, for then the number of counters in the row would be less than four, and they cannot both contain a 1 for then the number of counters would be more than 10.) There are only two ways that these three 1's (one for each row) can be arranged in the two columns: all three in one column, or two in one column and one in the other. In both cases, one column must total an odd number, making the initial position unsafe and thus guaranteeing a win for the machine if it plays first.

The following readers sent detailed analyses of the 4 × 4 Tac Tix game: Theodore Katsanis, Ralph Hinrichs, William Hall and C. D. Coltharp, Paul Darby, D. R. Horner, Alan McCoy, P. L. Rothenberg and A. A. Marks, Robert Caswell, Ralph Queen, Herman Gerber, Joe Greene, and Richard Dudley. No simple strategy was discovered, but there no longer is any doubt that the second player can always win on this board as well as on the 4 × 4 field with missing corner counters. It has been conjectured that on any square or rectangular board with at least one odd side, the first player can win by taking an entire center row on his first move, and that on fields with even sides, the second player has the win. These conjectures are, however, not yet established by proofs.

As things now stand, the ideal board for expert Tac-Ticians who have mastered the 4 × 4 seems to be the 6 × 6. It is small enough to keep the game from being long and tiresome, yet complex enough to make for an exciting, unpredictable game.

ANSWERS

The first Tac Tix problem can be won in several different ways: for example, take 9–10–11–12 or 4–8–12–16. The second problem is won by taking 9 or 10.

AFTERWORD, 1988

Nim belongs to a large class of what are called Nim-like or take-away games in which two players alternately remove counters from a specified set, according to specified rules. In standard play, the person taking the last counter wins. In reverse play, the last to take a counter loses. The classic pioneering paper on such games is "The

G-Values of Various Games," by Richard Guy and C. A. B. Smith in *Proceedings of the Cambridge Philosophical Society* 52 (July 1956): 514–526. The most extensive coverage of Nim-like games – scores of new and novel games are analyzed – is the four-volume *Winning Ways* by Elwyn Berlekamp, John Conway, and Richard Guy. Many of my later columns in *Scientific American* introduced Nim-like games, but I have limited my references in this chapter to traditional Nim.

Piet Hein's variant – I called it Tac Tix, but it later came to be known as Nimbi – has been examined by a number of computer programs without any general strategy emerging. The game need not, of course, be played with square patterns. They can be rectangular, triangular, hexagonal, or any other shape. In Denmark, in 1967, Piet Hein marketed a version played with an initial pattern of twelve counters in the form of an equilateral triangle from which the three corner counters have been removed. Aviezri Fraenkel and Hans Herda were able to prove that the second player can force a win in both standard and reverse play. They reported these results in "Never Rush to Be First in Playing Nimbi," *Mathematics Magazine* 53 (January 1980): 21–26.

POSTSCRIPT

There is now a vast literature on what are called Nim-like, take-away, or disjunctive games. Such games involve two players who take turns removing a counter, a cell, a line of a graph, or any other one of a finite set of defined objects. In standard play the last person to take away wins. In reverse or misère play, the last person to take loses. Usually, though not in the case of classic Nim, the reverse game is much more difficult to analyze.

The late graph theorist Frank Harary (he died in 2005) divided such games into what he called achievement and avoidance games. He published dozens of papers on original Nim-like games. Hopefully these papers, along with unpublished studies of such games, will someday be gathered into a book.

Many examples of interesting Nim-like games are discussed in books in this series to follow. See especially Chapter 16 of my

Book 6; Chapter 14 of Book 10; Chapter 14 of Book 12; and Chapter 8 of Book 13.

BIBLIOGRAPHY

"Nim, a Game with a Complete Mathematical Theory." Charles L. Bouton in *Annals of Mathematics*, ser. 2 (1901–1902): 35–39.

"A Generalization of the Game called Nim." S. H. Moore in *Annals of Mathematics* 11, ser. 2 (1910): 93–94.

"The Nimatron." E. U. Condon in *American Mathematical Monthly* 49 (May 1942): 330–332.

"A New System for Playing the Game of Nim." D. P. McIntyre in *American Mathematical Monthly* 49 (1942): 44–46.

"A Machine for Playing the Game Nim." Raymond Redheffer in *American Mathematical Monthly* 55 (June–July 1948): 343–350.

"Digital Computer Plays Nim." Herbert Koppel in *Electronics* (November 1952).

"The *G*-Values of Various Games." Richard K. Guy and Cedric A. B. Smith in *Proceedings of the Cambridge Philosophical Society* 52:2 (July 1956): 514–526.

"Disjunctive Games with the Last Player Losing." P. M. Grundy and Cedric A. B. Smith in *Proceedings of the Cambridge Philosophical Society* 52:2 (July 1956): 527–533.

"Matrix Nim." John C. Holladay in *American Mathematical Monthly* 65 (February 1958): 107–109.

"Compound Games with Counters." Cedric A. B. Smith in *Journal of Recreational Mathematics* 1 (April 1968): 66–77.

CHAPTER SIXTEEN

Left or Right?

THE RECENT "gay and wonderful discovery" (as Robert Oppenheimer called it) that fundamental particles of physics have a left and right "handedness" opens new continents of thought. Do all the fundamental particles in the universe have the same handedness? Will nature's ambidexterity someday be restored by the discovery that some galaxies are composed of antimatter – matter made up of particles that "go the other way," as Alice described the objects in her looking glass? Perhaps we can better understand these speculations if we approach them in a playful spirit.

Mirror reflections are so much a part of daily life that we feel we understand them thoroughly. Most people are nonetheless at a loss for words when they are asked: "Why does a mirror reverse left and right and not up and down?" The question is made more confusing by the fact that it is easy to construct mirrors that do not reverse left and right at all. Plato in his *Timaeus* and Lucretius in *On the Nature of Things* describe one such mirror, made by bending a rectangle of polished metal into the slightly concave form shown in the bottom left illustration of Figure 98. If you look into such a mirror, you will see your face as others see it. The reflection of a page of type may similarly be read without difficulty.

An even simpler way to make a mirror that does not reverse images is to place two mirrors, preferably without frames, at right angles to each other as shown in the bottom right illustration Figure 98. If you rotate this mirror (as well as the concave mirror) through 90 degrees, what happens to the image of your face? It turns upside down.

A symmetrical structure is one that remains unchanged when it is reflected in an ordinary mirror. It can be superposed on its

Figure 98. An ordinary mirror and its image (top) and two mirrors whose images are not reversed (bottom left and right). (Artist: Irving Geis.)

mirror image, whereas asymmetric structures cannot. The twin forms of all asymmetric objects are often distinguished by calling one "right" and the other "left." No amount of inspection or measurement of one will disclose a property not possessed by the other, yet the two are quite different. This sorely puzzled Immanuel Kant. "What can more resemble my hand," he wrote, "and be in all points more like, than its image in the looking glass? And yet I cannot put such a hand as I see in the glass in the place of its original."

This curious duality is found in structures with any number of dimensions, including those with more than three. A segment of a straight line, for example, is symmetrical along its one dimension; but if we consider a long segment followed by a short one, the pattern is asymmetric. Mirrored by a point on the linear dimension it becomes a short segment followed by a long one. If we think of printed words as symbols ordered in one dimension, then most words are asymmetric, though there are palindromic words like "radar" and "deified" which read the same both ways. There are even palindromic sentences. "Draw pupil's lip upward"; "A man, a plan, a canal – Panama!"; "Egad! A base tone denotes a bad age"; and Adam's first remark, "Madam, I'm Adam" (to which Eve appropriately replied, "Eve"). Poets occasionally make use of palindromic sound sequences. A good example is Robert Browning's well-known lyric "Meeting at Night" in which the rhyme scheme of *abccba* in each stanza was designed to suggest the movement of sea waves in the poem.

Melodies may similarly be regarded as tones ordered along the single dimension of time. During the fifteenth century, it was fashionable to construct palindromic canons in which the imitating melody was the other melody backward. Many composers (including Haydn, Bach, Beethoven, Hindemith, and Schönberg) have used the device for contrapuntal effects. Most melodies, however, grate on the ear in retrograde form.

Many amusing experiments in musical reflection can be performed with a tape recorder. Piano music played backward sounds like organ music because each tone begins faintly and swells in volume. Particularly weird effects may be obtained by playing music backward inside an echo chamber while recording it on another

Figure 99. The Chinese yin-yang symbol. (Artist: Irving Geis.)

tape. When the second tape is reversed, the notes regain their original order but the echoes precede the sounds.

Another type of musical reflection is produced by turning a player-piano roll around so that it plays forward but with high and low notes reversed – the inverted music a pianist would produce if he played in the normal manner on a looking-glass piano. The melody becomes unrecognizable, and there is an unexpected transposition of minor and major keys. This device was also used in Renaissance canons and in the counterpoint of later composers. The classic example is in Bach's *Die Kunst der Fuge*, in which the twelfth and thirteenth fugues may be inverted. Mozart once wrote a canon with a second melody that was the first one both backward and upside down, so that two players could read the same notes from opposite sides of the sheet!

Turning our attention to two-dimensional structures, we see that a configuration such as the Christian cross is symmetrical, whereas the yin-yang symbol, an ancient Chinese religious symbol (see Figure 99), is not. The dark and light areas, called Yin and Yang, symbolize all the fundamental dualities, including left-right and its combinatorial basis in even and odd numbers. The monad's pleasing asymmetry makes singularly appropriate the fact that it was two Chinese physicists (one of them named Yang!) who received the Nobel Prize in 1957 for their theoretical work that led to the overthrow of parity. Unlike music, all asymmetrical designs and pictures can be "flopped" (to use the graphic-arts term for "reflected") without losing esthetic value. In fact, Rembrandt once made a flopped etching of his famous *Descent from the Cross*. It has been suggested that left-to-right reading habits may have a subtle influence on a Westerner's reaction to a reflected picture, but if so, the influence seems to be slight.

Because most printed words form asymmetric patterns, reflections of printed matter are usually unreadable, but not always. If you look at a mirror reflection of the words "CHOICE QUALITY" on the side of a Camel-cigarette package, holding the pack so that its top points to your right, you will be startled by what you see. "QUALITY" is unreadable, but "CHOICE" is entirely unchanged! The reason of course is that "choice," when printed in capital letters, has an axis of symmetry and is therefore superposable on its mirror image by turning it upside down. Other words, like "TOMATO" and "TIMOTHY," are asymmetric when printed horizontally, but acquire an axis of symmetry when printed vertically.

When we consider familiar structures of three dimensions, we find that they are a pleasing mixture of symmetry and asymmetry. Most living forms are symmetrical in their outward appearance, with such notable exceptions as spiral shells, the pincers of the fiddler crab, the crossed bills of the crossbill, and the unilateral eyes of flatfish. Even behavior patterns are sometimes asymmetric; consider, for example, the counterclockwise gyrations of bats swarming out of Carlsbad Caverns. Most human-made objects are likewise symmetrical, though some that seem to be so prove to be asymmetric when inspected more closely – for instance, scissors, Moebius strips, hexaflexagons, and simple overhand knots. The two knots in Figure 100 have identical topological properties, yet one cannot be deformed into the other. Dice also have two distinct forms. There are two ways of placing spots on a die's faces so that the spots on opposite sides always total seven; one way is a mirror image of the other.

As Donald Knuth pointed out in a letter, there are two ways to place the six spots, two ways to place the three spots, and two ways to place the two spots. There are, therefore, $2 \times 2 \times 2 \times 2 = 16$ different forms of a die.

Since folding your arms is the same as tying them in an overhand knot, it follows that there are two distinct ways to fold arms, though we are so conditioned to one method that it is annoyingly difficult to execute its mirror twin. Fold your arms as you normally do, grasp the

Figure 100. Left- and right-handed Moebius strips (*top*), overhand knots (*middle*), and dice (*bottom*). (Artist: Irving Geis.)

two ends of a string, unfold your arms, and you will transfer the knot from your arms to the string. Repeat the experiment with your arms folded the other way and you get a knot that is a reflection of the first one. A fascinating (and unsolved) topological problem is to prove that a pair of mirror-image knots in a closed curve cannot be made to cancel each other by deforming the curve. No one has succeeded in doing it, though it is easy to push one knot into the other and form a square knot, which is symmetrical. If you do this with two knots of the same handedness, you get an asymmetric granny.

These are not trivial matters. Now that certain particles are known to be asymmetric in some as-yet-unknown spatial sense, physical theory will have to account for the fact that when a particle meets its antiparticle, the two annihilate each other and create symmetrical energy. Alice looked into her mirror and wondered if looking-glass milk was good to drink. For some time it has been known that such milk would not be digested, because the enzymes of the body, designed to act on left-handed molecules, could not cope with right-handed ones. Now it would seem that the situation might be a good deal worse. The recent parity experiments strongly suggest that a particle and its antiparticle are really nothing more than mirror-image forms of the same structure. If this is true, as most physicists suspect and hope, then any attempt by Alice to drink looking-glass milk would result in a violent explosion like the explosion of Dr. Edward Teller when (as dramatically described by Dr. Teller himself in *The New Yorker*, December 15, 1956) he shook hands with Dr. Edward Anti-Teller. It is safe to predict that physicists will be speculating right and left for a long time to come.

ADDENDUM

The question asked in the second paragraph of this chapter prompted the following letter from Dr. Robert D. Tschirgi and Dr. John Langdon Taylor, Jr., both of the department of physiology, School of Medicine, at the University of California Medical Center in Los Angeles.

Sirs:

The entertaining and provocative article on symmetry by Martin Gardner recalled for your readers the tantalizing question: "Why does a mirror reverse left and right and not up and down?" Despite the comprehensive descriptions of light paths and optical principles which are usually marshaled in answer to this query, there seems to be an even more fundamental basis, which, the writers of this letter propose, lies primarily within the province of psychophysiology.

Humans are superficially and grossly bilaterally symmetrical, but subjectively and behaviorally they are relatively asymmetrical. The very fact that we can distinguish our right from our left side implies an asymmetry of the perceiving system, as noted by Ernst Mach in 1900. We are thus, to a certain extent, an asymmetrical mind dwelling in a bilaterally symmetrical body, at least with respect to casual visual inspection of our external form. Here the term symmetry is used in an informational context, and indicates that the observer can make no distinction, other then sense, between two or more elements of his perceptive field. Of course by refining his observations he may gain information of other dissimilarities, at which time the system under consideration ceases to be symmetrical.

When we stand before a mirror, we see reflected a superficially bilaterally symmetrical structure, and we are misled by this apparent symmetry into treating the system as if ourselves and our reflection were identities rather than enantiomorphs (entities of opposite "handedness"). Therefore, by psychological projection, we seem to be able to rotate our body image 180 degrees in three-dimensional space around a vertical axis and to translate it a distance equal to twice the distance to the mirror, thereby achieving a coincidence between our body and its reflection. By this process we have imagined the identical central-nervous-system perceptive machinery which is in ourselves, rather than its enantiomorph, to exist within our mirror image. We are consequently led to the erroneous statement that when we move our right hand, our mirror image moves its left hand. If we, more correctly, imagine our enantiomorphic selves within our mirror image, then we realize that its definition of right and left would be reversed, and when we move our defined right hand, it moves its defined right hand. We must endow our reflection not with our own coordinate system, but with a mirror-image coordinate system. This can easily be illustrated by placing a paper bag over one hand and re-defining the major body axes as "head-feet," "front-back," and "hand-bag" (instead of right-left). Now stand before a mirror and observe that when you move head, mirror image moves head; when you move feet, mirror image moves feet; when you move hand, mirror image moves hand; and when you move bag, mirror image moves bag. What has become of right-left reversal? It has been dispelled, as the chimera it was, by the simple procedure of making our superficial structure obviously not bilaterally symmetrical. It is no longer possible to produce essential coincidence between ourselves and our mirror image by 180-degree rotation around our vertical axes, any more than around any other axis, and we recognize the enantiomorphic nature of our reflection.

To illustrate how the convention of rotation about a vertical axis imposes the concept of right-left mirror reversal on objects other than ourselves, consider a map of the U. S. oriented in the customary manner of North headward and East to the right. To observe the mirror image of this map, we invariably rotate the map around its North-South axis toward a mirror. This habit undoubtedly derives from the fact that most of our movements designed to inspect our environment involve rotation about our vertical axis. For example, if the map were fixed to a wall opposite a mirror, we would observe the map directly and then rotate ourselves about our vertical axis to view the map's reflection. In either case, East will now appear to our left, but North will remain up. If, however, we rotate the map around its East-West axis to face the mirror, or look at the reflection of the wall map by standing on our head, then East remains to our right, but North becomes footward. It now appears that the mirror has reversed top and bottom rather than right and left.

The only determined coordinate system is that which the observer imposes on his environment, and the axes can be adjusted so that the origin occurs at any point within the observer's perceptive space. When we describe the parts of an object relative to one another, we generally do so by adjusting our coordinate system so that the origin occurs within the object, and it thereby acquires top-bottom, front-back, and right-left axes corresponding to those of the observer. As objects rotate within this system, either through motion of the object or motion of the coordinate system (i.e., the observer), certain of the object's coordinate values will change sign. Rotation of an object around its vertical axis results in change of sign of right-left and front-back loci; around its right-left axis results in change of sign of front-back and top-bottom loci; and around its front-back axis results in change of sign of top-bottom and right-left loci. However, since the observer defines the coordinate system, rotation of the observer does not result in change of sign of the relative parts of the observer. Thus, if we look at our own reflection while standing on our head, we still erroneously interpret the mirror as reversing right and left, because in the process of inverting our body, we have inverted the coordinate system itself.

After this letter appeared in *Scientific American* (May 1958), the magazine received the following note from R. S. Wiener of Stamford, Connecticut:

Sirs:

After reading the interesting comments of Drs. Tschirgi and Taylor on the question "Why does a mirror reverse left and right and not up and down?" I decided to test some of their observations.

I tacked up a map (actually a chart of the Long Island Sound, Western Section) on the wall opposite the mirror over my dresser. Standing on my head on the floor in front of the mirror, I realized that I could not see all of my image. All I could see were two feet. The one that I recognized to be that which I usually term the left one was covering the section of the chart around Bridgeport, while the opposite foot was in the vicinity of the East River.

I then tried the experiment with a paper bag over the "left" foot. The bag was now hovering around Bridgeport. The experiment did not seem to be accomplishing very much, so I moved the dresser out of the room, took the mirror off the wall and put it on the floor, leaning it against the wall.

I again took my position on my head in front of the mirror. The image of the superficially bilaterally symmetric structure on its head with a bag over one foot was so frightening that I decided to drop the whole experiment.

POSTSCRIPT

Many years after my article "Is Nature Ambidextrous?" was published in 1952, and this chapter ran in *Scientific American*, I expanded the material into a book titled *The Ambidextrous Universe: Left, Right, and the Fall of Parity*. It was first published in 1964 by Basic Books, and revised in 1979 by Scribner's. A further revision, with new art, was issued by W. H. Freeman in 1990, retitled *The New Ambidextrous Universe: Symmetry and Asymmetry from Mirror Reflections to Superstrings*.

After the Freeman edition went out of print, I updated the book again in 2005 for a Dover paperback, with a new preface, and a raft of new footnotes including a long note on the absorption of superstrings into M-theory. I wanted to change the book's title to *The Asymmetric Universe*, but Dover preferred to keep the old one. Chapter 3 has several pages explaining exactly why a mirror seems to reverse left and right but not up and down.

BIBLIOGRAPHY

"On Symmetry." Ernst Mach in *Popular Scientific Lectures*. Open Court, 1895.

"Left or Right?" Martin Gardner in *Esquire*, February 1951. Reprinted in *Mathenauts: Tales of Mathematical Wonder*, ed. Rudy Rucker. Arbor House, 1987.

"Is Nature Ambidextrous?" Martin Gardner in *Philosophy and Phenomenological Research* 13 (December 1952): 200–211. Reprinted in *Order and Surprise*. Prometheus Books, 1983.

Symmetry. Hermann Weyl. Princeton University Press, 1952.

The New Ambidextrous Universe. Martin Gardner. Dover, 2005.

Index